ISBN 978-1-334-78628-0
PIBN 10548059

English
Français
Deutsche
Italiano
Español
Português

www.forgottenbooks.com

Mythology Photography **Fiction**
Fishing Christianity **Art** Cooking
Essays Buddhism Freemasonry
Medicine **Biology** Music **Ancient
Egypt** Evolution Carpentry Physics
Dance Geology **Mathematics** Fitness
Shakespeare **Folklore** Yoga Marketing
Confidence Immortality Biographies
Poetry **Psychology** Witchcraft
Electronics Chemistry History **Law**
Accounting **Philosophy** Anthropology
Alchemy Drama Quantum Mechanics
Atheism Sexual Health **Ancient History**
Entrepreneurship Languages Sport
Paleontology Needlework Islam
Metaphysics Investment Archaeology
Parenting Statistics Criminology
Motivational

CRITICAL REVIEW OF BIOLOGY
AND CONTROL OF OYSTER DRILLS
Urosalpinx and *Eupleura*

SPECIAL SCIENTIFIC REPORT–FISHERIES No. 148

UNITED STATES DEPARTMENT OF THE INTERIOR
FISH AND WILDLIFE SERVICE

EXPLANATORY NOTE

The series embodies results of investigations, usually of restricted scope, intended to aid or direct management or utilization practices and as guides for administrative or legislative action   it is issued in limited quantities for official use of Federal, State or cooperating agencies and in processed form for economy and to avoid delay in publication

United States Department of the Interior, Douglas McKay, Secretary
Fish and Wildlife Service, John L. Farley, Director

Critical Review of Biology and Control of Oyster Drills
UROSALPINX and EUPLEURA

By Melbourne Romaine Carriker
The Department of Zoology
The University of North Carolina
Chapel Hill, North Carolina

Special Scientific Report: Fisheries No. 148

Washington, D. C.
April, 1955

TABLE OF CONTENTS

List of Tables: Urosalpinx cinerea

CRITICAL REVIEW OF BIOLOGY AND CONTROL OF OYSTER DRILLS UROSALPINX AND EUPLEURA

INTRODUCTION

The oyster drill <u>Urosalpinx</u> <u>cinerea</u> (Say) is a small, slow-moving, highly successful, highly specialized, predatory marine snail inhabiting the coastal waters of North America and the British Isles. Because of its close association with and ravage of young oysters it has attracted the attention of oyster farmers for at least the last 100 years. Although the critical observations of oystermen and marine biologists during the last few decades have more accurately demonstrated the high concentration, broad distribution, and unusual destructiveness of this gastropod (Engle, pers. com.), it probably became a serious pest, concurrently with the development of widespread transplantation and cultivation of oysters. Glancy (1953) states that it is one of the worst enemies of the oyster, and blames the steady decline in oyster-meat production in the United States from 231 million pounds in 1910 to 77 million pounds in 1950 in large measure to the depredations of this drill. Stauber (1943) suggests that in addition to the evident destructiveness of the drill, a selective elimination of the faster growing, thinner shelled oysters may be taking place, resulting in an increasing proportion of thicker shelled, slower growing, less desirable oysters

It thus seems appropriate at this time critically to evaluate the information available on the biology and control of U. <u>cinerea</u>. The material presented in this synthesis has been assembled from numerous published and unpublished reports and personal communications through the generous cooperation of many persons, and brings into bold relief the many voids in our knowledge of this mollusk.

Further research on the morphology, physiology, and ecology of this gastropod will be facilitated by the fact that it is available in astronomic quantities in a wide variety of habitats in the coastal regions of two continents, is markedly adaptable to new environments, is tolerant of a broad range of environmental factors, is relatively immune to predation and parasitization by other organisms, and is easily maintained in the laboratory.

TAXONOMY

The oyster drill of western Atlantic coastal waters has been known by a
confusing list of scientific names: Fusus cinereus Say, Buccinum plicosum Menke,
Buccinum cinereum Gould, Urosalpinx cinerea (Say), and Urosalpinx cinereus Say.
Say (1822) first described this snail, giving it the name Fusus cinereus Say.
Later Stimpson (1865) created the genus Urosalpinx in the family Muricidae, included Fusus cinereus Say as type, and called it Urosalpinx cinerea (Say). The
latter is the proper form of the specific name; since "Urosalpinx" is feminine,
the adjectival specific name "cinerea" must agree in gender (Int. Code Zool. Nomen.,
Art. 14a, 1926, in Schenk and McMasters, 1948). The Family Muricidae has been
placed in the order Neogastropoda, Subclass Prosobranchia, Class Gastropoda
(Abbott, 1954).

Say's type drill is probably the small form, rather than the giant one,
since his measurement of the type is approximately 32 mm., and probably came
from Great Egg Harbor, New Jersey, where Say stated, that he had collected the
species (Pilsbry, pers. com.; Abbott, pers. com.), rather than from the eastern
shore of Maryland and Virginia where the giant form occurs (Henderson and Bartsch,
1915; Baker, 1951). Unfortunately Say's type specimen of "Fusus" cinereus has been
lost and is thus not available for study (Pilsbry, pers. com.).

All scientific names of mollusks used in this review are taken from the
nomenclature of Abbott (1954).

DISTRIBUTION
Fossil Distribution

U. cinerea is not a recent product of evolution. The genus was probably
initiated as early as the Eocene (J. Gardner, 1948) some 60 million years ago. The
oldest shells of the species have been taken in North Carolina (Richards, pers com.)
and in Maryland (Verrill and Smith, 1874) in Miocene deposits approximately 28
million years old. Shells of the species are more abundant in Pliocene deposits some
12 million years old of the Atlantic Coastal Plain in several localities in North
Carolina and in Florida (Richards, pers. com.; 1947) and also in South Carolina
(Verrill & Smith, 1874).

2

The species is reported as common along the Atlantic Coastal Plain in Pleistocene deposits approximately one million years old and in a range similar to that of modern descendants (Richards, pers. com.; Shimer & Shrock, 1944) Specifically it has been collected in marine deposits in Point Shirley and Nantucket, Massachusetts, and on Gardiners Island, New York (Verrill & Smith, 1874); in Barnegat, Beach Arlington, Peermont, Holliday Beach, Two Mile Beach, Cape May and Heislerville, New Jersey (Richards, 1933; 1944); in the coastal terraces of the Pamlico formation among marine fossils in Maryland, Virginia, North Carolina, and South Carolina, but not in Delaware (Richards, 1936); in the northeastern two-thirds of the coast and the mid-central west coast of Florida, but not on the southern or northwestern portions of this state (Richards, 1938); and in one locality in Louisiana, but not in Alabama, Mississippi, or Texas (Richards, 1939).

More recently U. cinerea was discovered in company with oyster shells in sediments of the Charles River estuary, Massachusetts, in excavated layers approximately 3,000 years old (F. Johnson, 1942).

Recent Vertical Distribution

Within its geographic limits the oyster drill occupies a relatively broad vertical range, extending in favorable salinities from the mid intertidal zone to unknown depths in the sea. Tryon (1873-74) reports its presence in maximum depths of 90 feet; Verrill and Smith (1874) and C. Johnson (1915), in 48 feet; and Dall (1889), in 60 feet. Chestnut (pers. com.) recently extended this range when he dredged a few specimens on the edge of the Gulf Stream, off Cape Lookout, North Carolina, in 120 feet of water, and observed evidence of drilling on bivalves accompanying the drills. As additional faunal studies are made in deeper water off our coastal areas the bathymetric range of the drill may well be extended.

Recent Horizontal Distribution
General

So far as the incomplete paleontological evidence discloses U. cinerea evolved somewhere along the middle Atlantic coast of the United States, and, before the Age of Man, spread intermittently over the range between Florida and Massachusetts. This dispersal probably occurred at an unhurried rate proportion ate, for the most part, to the sluggish locomotory behavior of the drill, and was abetted by the high degree of adaptability of the snail, the availability of transporting agents, and by accessibility of routes for movement to suitable new environments. Amelioration of physical, chemical, and biotic ecological factors previously obstruct ing its dispersion undoubtedly further fostered its distribution.

With the advent of civilized man and his subsequent exploitation of coastal shellfish resources, the distributional pattern of the oyster drill changed markedly and swiftly. Storer (1931) rightly points out that more changes in animal distribution have occurred within the past 150 years than in all previously recorded human history. Because of the close association of the drill with the oyster, it has become an ubiquitous uninvited passenger in the transplantation, cultivation, and harvesting operations introduced by man in the modern culture of oysters. Many of these methods inadvertently favor its wide dissemination and propagation. For example, unless carefully checked, the transplantation of infested oysters of any age from one region to another may introduce drills and their egg cases, and quite unintentionally live drills accompanying shell dredged for cultch may be shovelled overboard on other grounds. In this manner the extension of the range of this gastropod, unknowingly transplanted from its original haunts on barnacle and black mussel bottoms and natural oyster reefs, has kept steady pace with increased human utilization of new grounds for the culture of oysters (Adams, 1947; Bureau of Statistics New Jersey, 1902; Cole, 1942; Dall, 1921; Elsey, 1933; Engle, 1953; Federighi, 1931c; Galtsoff et al., 1937; Gibbs, pers. com.; Goode, 1884; Hanna, 1939; Ingersoll, 1881; Lindsay, pers. com.; Moore, 1898a, T. C. Nelson, 1922; Newcombe and Menzel, 1945; Orcutt, pers. com.; Orton and Winckworth, 1928; Orton, 1930; Rogers, 1951; Sherwood, 1931; Stauber, 1943; Storer, 1931; Townsend, 1893; Walter, 1910). Once established on oyster beds the drill soon attains high concentrations and persists there with phenomenal tenacity and success unless effectually controlled by man. The density of these drill populations probably fluctuates in large measure with the available food supply, drill breeding cycles, the abundance of enemies, and gradual or catastrophic changes in the physical and chemical environment.

Specific

Considerable information has been accumulated on the current geographic distribution of the oyster drill, and will be presented in detail with emphasis where available on the probable highways of dispersal. Just how far along these dispersion routes drills are able to survive and reproduce, particularly when the routes extend latitudinally, is still a matter of speculation. Much needed information on biological races of drills and the adaptability of these races to extremes in limiting environmental factors will certainly help in answering these important questions. Andrews (pers. com.) suggests that the distribution of drills by man is more important within local areas than over long coastal distances. Time may well support his conjecture, but one cannot fail to be impressed by the great quantities of living oysters, undoubtedly infested with drills at some stage of development, which have been transported long distances along the east coast of North America, to the west coast of North America, and to Great Britain. Such

4

transport along the east coast of North America has been a practice for at least the past 170 years; surely within this time acclimatization, quite likely a factor in the extension of the normal range of distribution, would have operated in an animal like the drill with a relatively short sexual cycle, particularly since stepwise transportation from bay to bay has occurred.

Eastern Coast of North America

Eastern Canada. In 1901 Whiteaves reported that Urosalpinx had extended its range northward into shallow sheltered comparatively warm waters of such areas as Passamaquoddy Bay and Minas Basin, Bay of Fundy; Halifax Harbor and Sable Island on the Atlantic coast of Nova Scotia; Magdalen Islands, Gulf of St. Lawrence; between Cape Breton and Prince Edward Islands, and Northumberland Strait; and on the northeast coast of New Brunswick to Carraquette Bay. He described its distribution as local and sparse. Chadwick (1905) found the drill quite rare in deep water in Northumberland Strait. Needler (1941) and Ingalls and Needler (1942) state that this snail is not generally distributed in Canadian Atlantic oyster areas but is abundant about the eastern part of Northumberland Strait in Malagash Basin, in Pugwash River, Tatamagouche Bay, Caribou Harbour, Merigamish, and in the vicinity of Charlottetown. They are abundant only where many oysters or mussels are found, and are destroying many oysters, but are said not to be serious to the oyster industry as a whole anywhere in Canada. Most recently Adams (1947) finds Urosalpinx in isolated restricted colonies on both sides of Northumberland Strait and in Minas Basin where they probably have existed for many years. In these two areas oysters occur only in Northumberland Strait where depredations are worst in Caribou Harbour and Malagash Basin.

Some drills may have been carried to eastern Canada when American oysters were transported to Malpeque Bay, P.E.I., for restocking after the local oysters were destroyed in 1915 and 1916 by poor conservation and disease (Sherwood, 1931). No records are available on possible earlier introductions.

Maine. The drill has been reported from only a few isolated sheltered bodies of water in this state. Verrill and Smith (1874) and C. Johnson (1915) found it in some of the warmer shallow branches of Casco Bay, especially in the upper end of Quahog Bay. These may well have been introduced by man, since the importation of oysters from Virginia to Portland, Maine, commenced about 1840, and surplus from each cargo was shipped to Casco Bay and left on the flats for summer storage (Ingersoll, 1881). In 1895 Wentworth reported the drill as common in Damariscotta and Newcastle. More recently Galtsoff and Chipman (1940) in an exploration of the upper Damariscotta River found a few live Urosalpinx in the area where extensive oyster grounds existed during precolonial time. It is also

5

possible that the island-like areas of distribution of Urosalpinx in both Maine and Canada represent marine relicts, isolated survivors of a former widely distributed population of drills existing under more favorable conditions of temperature.

Massachusetts. A review of the rate of importation of living oysters into this state, as reported by Gould (1841) and by Ingersoll (1881), illustrates the degree to which U. cinerea was probably introduced from southern waters. Shortly after 1780 when the local supply of oysters was exhausted the supply for the markets of all the large towns was obtained from the south. By 1820, 12,000 to 14,000 bushels were brought annually to Wellfleet, Cape Cod, at first from Buzzards Bay and Narragansett Bay, later from Connecticut, and finally from New York and New Jersey. By 1841 some 40,000 bushels were being imported annually from New York, New Jersey, and to a lesser extent from Delaware Bay and Chesapeake Bay. The oysters were planted principally in Wellfleet, and in small quantities in Boston Harbor and other seaports, where they were left in the water to grow 7 to 9 months.

The distribution of the drill in the state is given imperfectly by a number of investigators in the last century, but the relationship between the exotic and the native drills, if any, and the influence of the former on the distribution of the species is unknown. In 1841 Gould collected Urosalpinx in the bays and inlets about Nantucket, New Bedford, and occasionally in Boston Harbor. In 1870 he extended the range to Vineyard Sound, Lynn Harbor, St. Simons Island, and Georgia. In 1874 Verrill and Smith noted that the drill was abundant in Vineyard Sound and Buzzards Bay and that its range then extended northward to Massachusetts Bay. Ingersoll in 1881 wrote that the most destructive enemy of the oyster was the drill, and that some beds, particularly those on hard bottom in Wareham, were completely destroyed by them. He reported that in Taunton River the drill was becoming more numerous and troublesome and destroyed 9/10 of the oyster seed between Somerset and Assonet. In 1884 Goode observed that the drill was present in such natural haunts as the rocky shores of Buzzards Bay and was hard to eradicate. C. Johnson in 1915 collected them in Lynn, Cohasset, Vineyard Sound, Buzzards Bay, and Nantucket.

Galtsoff (pers. com.) reports that in the Cape Cod region at the present time Urosalpinx are extremely abundant on the rocks in Woods Hole Harbor and in the adjacent portions of Buzzards Bay. They can also be found widely distributed in the inshore waters of the Cape and are abundant in the tidal streams and inlets in the upper part of Buzzards Bay, especially in Onset and Wareham regions, in Oyster River near Chatham, and in Wellfleet Harbor. Oyster River is a small tidal stream where the oyster industry of Cape Cod is being continued by a group of oystermen, but against considerable odds. Because of the abundance

6

of drills, oysters cannot be raised from seed here. Instead adult oysters are shipped each spring from Long Island and New Jersey to be sold the following fall and winter. Setting of oysters in Oyster River is fair, but drills rapidly destroy the sets.

Rhode Island. Gibbs (pers. com.) writes that Mr. Reynolds, an oysterman active during the period 1870-1900 in the Wickford, Rhode Island, area, first noticed drills in this state in 1870 and believes they were transported on oyster seed from Chesapeake Bay to the Conimicut Point oyster beds. Ingersoll wrote in 1881 that the drill occurred abundantly in many parts of Narragansett Bay in that period and rapidly destroyed oyster seed and younger oysters. In 1888 Rathbun found many drills in Narragansett Bay and Providence River, and noted that Urosalpinx was so abundant and so destructive between Gaspe Point and Pawtuxent Beach that owners relinquished claims to oyster beds; and that Fields Point and Bullocks Cove, which were formerly the most productive oyster beds in the river, were badly overrun by drills. Rowe (1894) estimated that the damage done by Urosalpinx in New England as early as 1894 was approximately one million dollars. In 1902 Carpenter also found drills extremely abundant in Narragansett Bay and remarked that they destroyed a great many oysters in a short time. C. Johnson in 1915 found the drill well distributed in Narragansett Bay and Watch Hill; and in 1937 Galtsoff et al. wrote that this pest was very abundant on the rocky shores of New England.

Connecticut. This state received its share of imported oysters from southern waters and in this way may have added to its native drill fauna. According to Ingersoll (1881) and Collins (1891) the importation of oysters, principally to the New Haven area, from the lower Chesapeake Bay and tributaries began between 1830 and 1840 in sailing vessels, and reached a maximum between 1855 and 1860. In 1891 Collins found Urosalpinx very troublesome to oyster planters, particularly destructive to small oysters, and most abundant in New Haven Harbor where "they apparently increase in numbers and destructive power each year". Later C. Johnson (1915) observed drills commonly distributed in New Haven and Stratford, and Jacot (1924) found them along various beaches near Bridgeport.

New York. Ingersoll (1881) has given an excellent description of the abundance of oysters in New York Bay and vicinity at the time white man first colonized these shores. Oysters grew in abundance over much of New York Bay, and in the lower reaches of the Shrewsbury, Raritan, Passaic, Hackensack, Hudson, and East Rivers. Approximately 50 miles of magnificent oyster reefs extended up the Hudson River from Sandy Hook. By 1810 the oyster resources in this area were apparently depleted by man. In combination with these local changes, it is highly probable that drills were inadvertently imported from the south on oysters sailed in for planting from Chesapeake Bay as early as 1816.

The first record of drill distribution in this state was given by DeKay in 1843. He found Urosalpinx very common on the eastern coast and very destructive to oysters. Ingersoll (1881) wrote that in 1878 the drill proved a great nuisance about East Point, injuring many oyster beds beyond repair. He was told that in Great South Bay, Long Island, where formerly no drills were reported, oyster grounds were currently overrun with them. Rathbun (1888) found the drill very persistent and destructive in New York waters. J. Nelson (1893) observed that drills were more abundant in Long Island Sound than in New Jersey, and thought that this might be explained by the fact that in Long Island Sound oyster seed had been raised by "shelling" for some years So far as the writer knows this is the first recorded suggestion that Urosalpinx is benefited by oyster management practices; a fact so well substantiated in the years to follow. A Gardner (1896) dreding in various parts of Long Island Sound found drills abundant along the shores of Port Chester Harbor, and less numerous in Oyster Bay and Lloyds Harbor. Moore (1898a) observed that the deep water beds of Long Island Sound were suffering increasingly from the drill, and supposed that it might be accounted for by the use of oyster seed from the drill infested beds in the less saline inshore grounds. In 1902 drills were particularly plentiful about the waters of Staten Island Sound (Bureau of Statistics New Jersey, 1902). Weeks (1907) found the drill common at Northport, Long Island. Galtsoff et al. (1937) found them very abundant on oyster bottoms of Great South Bay, Long Island; and Engle (1953) states that they are widespread throughout Long Island Sound.

New Jersey. The earliest record of the presence of Urosalpinx in this state is that by Say in 1822 who frequently found it along the coast. This record, though lacking in detail, suggests that the drill existed in New Jersey before the importation of oysters and the possible introduction of drills from other areas.

The principal enemy of the oyster in the Delaware Bay estuary in 1881 according to Ingersoll was the oyster drill which overran oyster bottoms on which it was not reported formerly. It was again mentioned as a serious pest in 1887 (Stauber, 1943). Many oysters were imported annually from Chesapeake Bay to the west shore of Delaware Bay for planting (Ingersoll, 1881) and probably introduced Urosalpinx from these waters. In the summer of 1910 Urosalpinx was said to be very destructive on the public oyster beds (Pope, 1910-11; Moore, 1911). In the next decade T. C. Nelson (1923) estimated the loss to New Jersey due to this pest in excess of one million dollars annually. The seriousness of the depredations stimulated considerable research on Urosalpinx between 1930 and 1942. This research was begun by J. R. Nelson, who was followed by I. W. Sizer, and he by J. B. Engle, and was greatly extended by L. A. Stauber. In 1937 Galtsoff et al., as a result of some of these studies.

8

reported the drill as abundant. Stauber (1943) in his 7 years of investigations in this estuary found Urosalpinx at all points over the oyster planting areas, and on the lower portion of the natural beds upstream.

The drill has also been found in abundance in the estuaries along the east coast of the state. Wood and Wood (1927) found it along Seven Mile Beach, Cape May. An early record by Ford (1889) lists it as occurring south of Brigantine Island, near Atlantic City. T. C. Nelson (1923) reports it as occurring abundantly in Little Egg Harbor and Barnegat Bay. During the last 15 years the writer has found this snail in sizeable concentrations in Great Bay, Little Egg Harbor, Barnegat Bay, and Shark River.

Scattered information is also available on the earlier history of the drill in Raritan Bay and its tributaries (Bureau of Statistics New Jersey, 1902), where it was considered particularly plentiful in 1902. The report states that the drill first appeared in the Shrewsbury River area in 1892, imported from Connecticut in a shipment of 25,000 bushels of oysters that were planted in local waters. These oysters and many local oysters were entirely destroyed by the drills which in one season were said to multiply to such an extent as to "cover the river bottom". Destruction of oysters by drills continued through 1895 and 1900 but in the spring of 1901 unusual spring freshets virtually exterminated the drills and but little loss of oysters was experienced. It is difficult to believe on the basis of related data on drill distribution presented in this review and from Ingersoll's (1881) statement that drills were abundant on oyster beds in nearby Keyport, that Urosalpinx was first introduced into this area in the 1892 shipment. It is more likely that favorable local conditions, in addition to the introduction of additional food and drills, stimulated an unusual reproduction and survival of both local and imported drills.

Maryland. Say (1822) frequently found Urosalpinx on the eastern shore of Maryland, and Ingersoll (1881) noticed that nearly every dredge haul in the lower Chesapeake Bay waters in Maryland brought up drills, while the Potomac seemed to be the least infested. Engle (1953) finds the mollusk in Chincoteague Bay and other coastal bays where salinities are high, but of non-commercial significance in Chesapeake Bay waters north of the Potomac River where salinities are low over much of the year, and notes (pers. com.) that lower Tangier Sound supports a drill population that fluctuates in numbers and position according to the fall of salinities during dry and wet years.

Virginia. Rogers (1951) suggests that the drill originated in Chesapeake Bay and from this locality was transported north and south. The history of oyster culture strongly suggests that Urosalpinx was exported on oysters from

9

this estuary during the occupancy of the eastern coast by civilized man. but according to the fossil record the drill was well distributed along the coast before this time. Unintentionally man has accelerated the mixing and dispersal of the species.

Urosalpinx has been reported from the eastern shore of Virginia since the last century (Ryder, 1883; Federighi, 1931c; Newcombe & Menzel, 1945). But lower Chesapeake Bay and its tributaries also nurtured a fair population of U. cinerea at least as early as the latter half of the last century, contrary to the reports of Federighi (1931c) and Newcombe and Menzel (1945) who suggest that early in the present century the drill was limited principally to the eastern shore and was spread into the Chesapeake on oysters. Uhler as early as 1878 described the drill as common on the rocks in the vicinity of Fort Wool, and Ingersoll (1881) and Goode (1884) both wrote of the abundance of drills over much of the lower Chesapeake. Ryder (1883) found them more or less abundant in all waters in which oyster culture was practiced. Rathbun (1888) reported them in the Chesapeake, but states unrealistically that they gave little trouble, while Moore (1898a) wrote that drills were the most destructive enemies of the oyster in the Chesapeake and adjoining areas. Federighi (1931c) described Urosalpinx as occurring over the whole Hampton Roads area and more abundantly on planted areas than on natural bottoms. He concluded that the greatest infestation of drills obtained in Chesapeake Bay and in the coastal waters of the northern states commensurate with the intense culture of oysters in these waters, while to the south where oyster culture was rare, the drill was insignificant. Galtsoff et al. in 1937 reported that drills were very abundant on the eastern shore and in the lower Chesapeake Bay and the lower portions of its tributaries where salinities remain high. Newcombe and Menzel (1945) found the drill unevenly distributed over most of the bottoms of the bay and rivers with a salt content above 15 o/oo. Engle (1953) confirmed the abundance and ubiquitousness of this muricid in this area and considers it the principal enemy of the oyster. The giant form of Urosalpinx from the eastern shore of Maryland and Virginia has never been reported elsewhere. In view of many possible avenues for dispersion by man, it would seem that this race has evolved and remained within a narrow set of ecological limits found only in its present habitat. Experimental transplantation to other regions has not been attempted; such studies may shed light on the nature of the factors which enforce this restricted isolation.

Southeastern States. Very little is known of the distribution and abundance of U. cinerea in this area as a whole. Since in the late 19th. century oyster culture was carried out only to a limited degree (Ingersoll, 1881) there was little opportunity to observe the activity of this snail, and the propagation of the drill was not benefited by the extensive cultural practices which augmented the drill populations of northern waters. Moore wrote that in 1898 (1898b) the drill was

10

practically harmless in the southeast. Further zoogeographic research will undoubtedly show this gastropod to be more abundant and widespread here than it now appears.

North Carolina. Hackney (1944) lists drills as extremely common around Ostrea. Pearse (1951) found them in the Bearfort Estuary. Chestnut and Fahy (1953) and Chestnut (pers. com.) in 1954 find U. cinerea common in all the sounds from the South Carolina line to Pamlico Sound, especially near the inlets. In Lockwood Folly River, Brunswick County, three sample plots yielded 13 Urosalpinx per square yard, and in Saucepan Creek, in the same county, a concentration ranging from 9 to 106 drills per square yard. A single cluster of serpulid tubes collected in New River in April yielded 21 Urosalpinx. From the piling at the Institute of Fisheries Research pier in Bogue Sound Chestnut has taken as many as 120 drills above the low water mark from a single piling. The writer has also found high concentrations of drills in the vicinity of the pier. Chestnut finds the drills present in some abundance around Pivers Island and at Cape Lookout, and plentiful in the upper part of Core Sound, where they apparently appeared when Drum Inlet broke through during the storm of 1953. In Pamlico Sound itself there are few Urosalpinx except in the proximity of the inlets. Chestnut reports only an occasional drill in the western half of this sound in the last 6 years. He has also dredged them off Cape Lookout on the edge of the Gulf Stream

South Carolina. There are conflicting reports as to the early distribution of Urosalpinx here. In 1890 Dean said it occurred rarely, and in 1913 Mazyck stated it was abundant. Galtsoff et al. (1937) found them only sparingly. The most recent detailed and accurate information comes from Lunz (pers. com.). In a preliminary survey of the coastal waters from Santee River southward to the Savannah River in 1935, he found U. cinerea all along this coast, but approximately twice as abundant in the northern half. In 1938 he observed more drills in Harbor River than in any other river of comparable size in the state, concentrations up to 36 large drills per square yard at the mouth of the river. During the last few years he has noticed an unusual increase in the abundance of the drill at the laboratory dock at Bears Bluff, while in other areas in the state he finds Urosalpinx widely but irregularly distributed and relatively scarce, certainly as compared to the concentrations reported by others in Chesapeake Bay. No explanation is known for the spotty distribution. Lunz believes the general paucity of this mollusk may be explained by the practice in his state of growing the vast majority of oysters intertidally. Andrews (pers. com.) suggests that the presence of extensive areas of soft mud may further limit distribution of the drills.

Georgia. Galtsoff et al. (1937) report that Urosalpinx occurs only sparingly here. More recently the staff at the Marine Biology Laboratory of the University of Georgia on Sapelo Island report that the drill is very abundant below the low tide line in the immediate vicinity of the island. Nothing is known about the distribution or abundance of the drill elsewhere in the state (Pomeroy, per.com.).

Florida. As early as 1874 Verrill and Smith stated that Urosalpinx was present along northeastern Florida and on the west coast in Tampa Bay. Between 1881 and 1888 C. Johnson (1890) commonly found it on oysters at St. Augustine. Ruge (1898) wrote that it was not found in this state; he did not list specific areas and hence may have been speaking of regions other than those from which the drill has been reported.

Northern Coast of the Gulf of Mexico. Apparently U. cinerea does not occur here. Humm (pers. com.) in the Alligator Harbor area, Tallahassee, and Butler (pers. com.) in the Pensacola region, have never encountered it. Nor has A. E. Hopkins (pers. com.) seen it along the Gulf from Apalachicola, Florida, to Corpus Christi, Texas. Hedgpeth (1953) also notes that this muricid is unknown in the living fauna of the area.

Bermuda. Federighi (1931c) erroneously extended the modern range of U. cinerea to this island on the basis of a reference by Arey and Crozier (1919). Close examination of this paper reveals only that chitons were drilled by oyster drills whose scientific name was not given. Verrill (1902), among others, makes no mention of U. cinerea either as a native or as an introduced species on the island. And Haas (pers. com.) who studied the mollusk fauna of Bermuda for some time, never found U. cinerea there, nor knows of any earlier or later record of the species in the Bermuda group. The writer concludes that the drill of Arey and Crozier was another species. The mention by Galtsoff et al. (1937) of U. cinerea in Bermuda (no reference is cited) is probably taken from Federighi's (1931c) report.

Western Coast of North America

There may be some question as to the exact role which man has played in the dispersion of U. cinerea along the east coast of North America principally because of the fossil distribution of the species in this region. However the transportation to and subsequent distribution of this snail along the western coast of North America is a clear demonstration of unwitting human collaboration in the spread of an undesirable species. The introduction of Urosalpinx to English waters is an equally striking example.

California. Soon after the opening of direct rail communication between the east and the west coast, the firm of A. Booth & Co. transported three carloads of large live eastern oysters to San Francisco. This is reported as the first shipment of live oysters from the Atlantic coast. Walter (1910) states that these particular oysters were harvested in Princess Bay, Staten Island, New York, in 1871; Collins (1892) gives the date as approximately 1869; and Hanna (1939), about 1870. This shipment overstocked the San Francisco market, and surplus oysters were planted in San Francisco Bay. The oysters grew well and importation continued. At one time transportation of young oysters reached enormous proportions; and oysters were planted in many parts of San Francisco Bay as well as in other inlets along the coast where they developed readily to marketable size (Hanna, 1939). U .cinerea are thought to have been introduced, if not in this first shipment, certainly on subsequent ones. (Townsend, 1893; Walter, 1910; Dall, 1921; Hanna, 1939). It was first recognized in California waters by Town send (1893) who wrote that it was becoming troublesome at this time on oyster beds in San Francisco Bay, particularly in the southern part of the bay where drills were most abundant. Stearns (1900) thought that the oyster drill was discovered on the oyster beds near Belmont on the westerly shore of San Francisco Bay as long ago as 1889 by Townsend; however Townsend's own report (1893) implies a much earlier date. Hanna (1939) states that Urosalpinx was collected on the Alameda flats in 1898, on oyster beds near Belmont in 1889, and near Redwood City in 1899. Smith in 1907 noted that the oyster drill had become very abundant and several years earlier was reported to be destroying oysters at a rate of 30 thousand dollars annually. At this time oyster seed was still brought yearly from New York and vicinity for planting.

Orcutt (pers. com.) has kindly provided data on the present distribution of U. cinerea in California, where its distribution coincides in time and place with major plantings of oysters from the Atlantic coast. The practice of planting eastern seed oysters in San Francisco Bay established in 1871 continued until 1900. From 1900 to 1932 half grown oysters were utilized. Now Urosalpinx is found generally throughout South San Francisco Bay. In Tomales Bay there is an area which has been used to hold full grown eastern oysters for the San Francisco market since 1875. These grounds, approximately 500 acres in size, are heavily infested with the oyster drill. In Arcata Bay (North Humboldt Bay) in northern California there is another area, of approximately 200 acres, which has been diked and over which eastern oysters were planted in 1910 and 1911 and again in 1935 and 1936, which also supports U. cinerea. It is important to note that areas of oyster culture in California where other than imported eastern oysters have been cultured, do not appear to support the eastern drill.

On the other hand there have been small importations of eastern oysters planted in other locations in California waters in which the drill has not been reported to date. This suggests that Urosalpinx may not be able to adjust to these habitats, or that insufficient drills were imported to colonize the areas. Since the drill is able to establish itself in waters in which the eastern oyster does not reproduce, as in most waters of the west coast, the latter is the more likely explanation.

13

Oregon. The failure of Urosalpinx to establish itself is strikingly illustrated in this state. Marriage (pers. com.) writes that Oregon's coastal waters are free of the eastern drill in spite of shipments of eastern oysters to this state in the middle and late 1800's. No recent shipments of oysters are reported. No reason for the absence of Urosalpinx here is available, except perhaps that insufficient drills were imported, or that ecological barriers prevented reproduction.

Washington. In 1906, 95 carloads of eastern oysters were introduced to Willapa Harbor (Elsey, 1933), and in 1907 Smith reported that the Bureau of Fisheries also planted 80 barrels there. Lindsay (pers. com.) writes that at the present time this snail occurs sparsely in Samish Bay, Padilla Bay, Rocky Bay-Case Inlet, Oakland Bay, Oyster Bay-Totten Inlet, Mud Bay-Eld Inlet, Nisqually Flats, Frinnon Flats, and Willapa Harbor. It is likely that the drill entered Samish Bay, Padilla Bay, Oyster Bay, and Willapa Harbor on direct trans plantation of year-old oyster seed from the Atlantic coast, and on transplantations from San Francisco Bay to all of these bays except Padilla Bay. Infestations present in Rocky Bay, Oakland Bay, and the Nisqually Flats are definitely traceable to transplantations from other bays. He states that evidence of serious damage to oysters by this drill in Washington is not indicated to date. Chapman and Banner (1949), for example, reported that in Mud Bay where U. cinerea is present and the Japanese drill is absent, the total number of oysters drilled on all beds was less than 1%. Lindsay suggests that generally the habitats found in Puget Sound are not particularly favorable to the survival of U. cinerea and it does not occur there in large concentrations.

Western Canada. In 1906, three or four carloads of eastern oysters were planted in Boundary Bay and Esquimalt Harbour, British Columbia. Importations into this region at first consisted of seed oysters, but because of high mortalities, three to four year old oysters were transplanted. Considerable mortality occurred among these also, so by 1912 importations diminished considerably, and by 1933 only two to three carloads were imported annually to British Columbia. As a result of these importations U. cinerea occurred plentifully in Boundary Bay and at Crescent, and less abundantly in Ladysmith Harbour by the early 1930's (Sherwood, 1931; Elsey, 1933).

Great Britain
The establishment of U. cinerea in English waters represents another remarkable extension of the range of this hardy animal by man. A few early records shed some light on the time and means of its introduction. Ingersoll (1881) writes in his highly informative report that about 1871 a New York oyster dealer began the exportation of American oysters into English markets where they sold

14

well and a brisk export trade developed. Liverpool was the principal receiving port for Great Britain. The quantity of oysters sent each week, though not large, was more than could be disposed of before the next shipment so surplusage was planted in local waters to be drawn upon as required. In addition thousands of barrels of younger oysters were exported to be held in English waters from one to three years. Unfortunately Ingersoll makes no mention of what coastal areas were utilized for holding the American oysters. Stauber (pers. com.) adds that years ago (dates not given) an oyster dealer, Mr. Beach, also used to make regular and extensive shipments of oysters to Britain from waters north of New Jersey. Numerous drills were undoubtedly introduced with these early shipments.

Korringa (pers. com.) writes that it is probable that U. cinerea was first introduced into English waters on a large experimental consignment of American oysters in connection with the great Fisheries Exhibition in London in 1883. These oysters were relaid in east coast waters where the drill probably became established. Orton (1909) notes that Crepidula fornicata was first introduced into England from America on American oysters about 1880. According to Orton and Winckworth (1928) there can be no doubt that Urosalpinx was introduced in the same way and probably about the same time as C. fornicata.

The earliest authentic record of the occurrence of U. cinerea in England is that of Orton (1930) who found it among animals preserved in 1920. But Orton points out that this drill was no doubt present in English waters for many years and remained undetected until 1927 when experiments were being conducted with native English drills (Ocenebra and Nucella).

Nor is there doubt that Urosalpinx can survive the passage across the Atlantic in the holds of ships. In 1939 Cole (1942) was informed that several living drills had been found among American oysters received by an east coast oyster merchant. He concludes that the possibility of fresh introductions of drills will exist as long as American oysters are imported. The continued importation of the Japanese oyster drill, Tritonalia japonica, to the west coast of North America on Pacific oyster seed from Japan illustrates this danger in a closely related drill (Chapman & Banner, 1949). This danger is more acute than previously anticipated. Woelke (1954) has shown that approximately 85% unhatched Japanese drills of various stages of development in the egg case can survive shipment from Japan to the United States out of water in the holds of ships among seed oysters for as long as 22 days.

According to Cole's (1942) best estimate, U. cinerea is not found outside of Essex and Kent, although no intensive research has been conducted for it in other areas. The two main centers of distribution appear to have been Brightlingsea

15

and West Mersea, Essex, where American oysters have been laid down for many years. The eastern American drill is very abundant in the River Blackwater and the River Colne; it occurs abundantly in the River Crouch and even more abundantly in the River Roach. It seems likely that Urosalpinx was introduced into the Roach-Crouch River system on Littorina or on oysters shortly before 1934. On the Kent coast Urosalpinx occurs sparingly at the mouth of the River Swale, which is apparently an unfavorable habitat since it never has become abundant there in spite of numerous importations of oysters from Essex. Cole points out that some habitats are more favorable for the survival of Urosalpinx than others, and thus may not necessarily become established in all areas where introduced. Examples of this have already been given for the west coast of North America.

Other Areas

According to available records Urosalpinx cinerea is not found beyond the geographic range already described for it in the previous sections. Further zoo-geographic research and continued transportation of living shellfish by man will undoubtedly extend its recorded range

U. cinerea is lacking in collections made in Holland (Korringa, pers. com.) and in the Netherlands, Belgium, and France (Jutting, pers. com.). Korringa believes that because of the shortage of oysters in England and the consequent ship ment of oysters there from Holland and France the danger of introduction of the American drill into Europe is minimized.

No American drills have ever been found in South Africa (Korringa, pers. com.; Day, pers. com.) even though experimental shipments of oysters from Europe are now being cultured in Knysna. Day believes Urosalpinx may be introduced on these oysters, but Korringa thinks this very improbable since oysters are being shipped from Arcachon or Brittany where Urosalpinx does not occur

Nor has mention been found of the occurrence of U. cinerea in Australia (Roughley, 1925) or in Japan (Cahn, 1950). Pilsbry (1895) lists the genus Urosal-pinx as occurring in Japan, but not the species U. cinerea.

Temporal Distribution

A few scattered reports faintly suggest that the size of drill populations may fluctuate over long periods of time. Ingersoll (1881) remarks that the disappearance of the drill from certain restricted localities for a long time is unexplained, and cites an instance in 1878 when the drill was very destructive in the waters around East Point, New York, only to practically disappear after that. Dall (1907) states that drills once numerous on planted oyster beds in

16

San Francisco Bay, were not in evidence there later. Higgins (1940) observed that during the last few years the oyster drill has become very abundant and destructive in the waters of Long Island Sound. Stauber (1943) observed relatively high concentrations of egg cases in Delaware Bay in 1937, and relatively low densities during the following three years.

Whether such fluctuations are haphazard, or cyclical and predictable in nature, can only be determined from careful quantitative studies carried out over a period of decades. Should such drill population trends prove to be predictable, this information would be of great value in control.

ABUNDANCE

A limited number of reports on the concentration of U. cinerea in a variety of habitats indicate that during the summer months the drill tends to occur most densely on intertidal reefs, piling, rock surfaces, and on oysters on oyster bottoms, undoubtedly a reflection of its negatively geotactic response at these temperatures. On subtidal oyster grounds in Delaware Bay, New Jersey, an average of approximately five drills per square meter was removed on baited drill traps in 1936 from a 20 acre plot (Stauber, 1943). Since this method of capture does not remove all of the drills, actual concentrations were probably higher. Mistakidis (1951) obtained a maximum density of 6 drills and an average density of about two drills per square meter on a subtidal oyster ground in the River Crouch and Roach, England. On a relatively flat piece of intertidal bottom in Little Egg Harbor, New Jersey, T. C. Nelson (1922) counted 29 drills per square meter. In 1953 the writer encountered concentrations of adult drills as high as 344 per square meter on the vertical intertidal surfaces of encrusted rocks off the west end of Gardiners Island, New York. The highest densities so far reported are those recorded by Stauber (1943). He obtainted counts ranging from 237 to 947 drills per square meter on an intertidal oyster reef several hundred square meters in size growing on a slag pile surrounded by sand in Delaware Bay.

The briefness of this section is an accurate reflection of the paucity of quantitative data available on the density of drills. Further information, but mostly of a qualitative nature, is presented in the section on "Distribution"

FORM AND FUNCTION
General

A large part of the information available on the anatomy of the oyster drill has been reported by the writer (1943). This, although dealing with many of the organ systems of the animal, emphasizes the minute anatomy of the anterior portion of the alimentary canal. Much of this anatomy is so detailed that it seems appropriate, as a basis for a better understanding of the ecological portions of this review, to include here a simplified condensation.

The shell of U. cinerea has been illustrated in a number of publications (J. R. Nelson, 1931, these drills were collected in Hueys Creek, Little Egg Harbor, New Jersey; Federighi, 1931c; Galtsoff et al., 1937; Cole, 1942; Carriker 1943). It is dextral, thick, and solid and thus affords a considerable degree of protection, spirals conically, and bears spiral striations and longitudinal ribbing. A short siphonal canal extends forward from the small aperture of the shell. In a series of five tests Sizer (1936) determined that shell material of adults possesses a salt content of 1.30 o/oo. When the animal is retracted within the shell the aperture is tightly closed by a strong chitinous operculum which is borne upon the rear upper surface of the foot. Just how long a drill may remain tightly sealed within its shell under a variety of conditions is not known. J. R. Nelson (1931) notes that at summer temperatures drills have been known to remain alive out of water for several days, presumably in the shade.

When normally expanded the exposed soft parts of the drill, consisting of a small foot, head, and tentacles, extend but a short distance outward from the shell. A pair of slender tapering retractile tentacles, nearly united at their bases, and each bearing a jet black eye along the mid outer side, arises on the front of the head and points forward. In males a long tapering "C-shaped" penis, about the length of an extended tentacle, lies on the right side of the head, and because it is hidden under the shell is rarely visible.

A false mouth lies just below the base of the tentacles. The true mouth is found at the tip of a long trunk-like proboscis which is normally housed within the head region. The proboscis is everted through the false mouth when the snail is drilling or feeding. Drilling is facilitated by two accessory structures. The first consists of a tube formed by the inward overlapping of the lateral ridges of the front part of the foot. The proboscis moves within this fleshy cylinder, receiving support and protection therefrom when everted. The second structure is the accessory proboscis, a gland lying in a cavity in the mid anterior ventral portion of the foot. The opening to this cavity is very difficult to see macroscopically except when the gland is functioning. In females an egg case pouch lies directly behind the accessory proboscis and is visible externally as an oval constricted depression.

All external surfaces of the drill, especially the ventral surface of the foot, are covered by a thin sheet of epithelium which secretes mucus and is covered with

18

cilia. The cilia maintain the surfaces free of sediment, understandably a major problem with gastropods which live on the bottom in water that contains varying quantities of sediment in suspension

The internal surfaces of the shell are blanketed by a sheet of tissue, the mantle, which becomes thickened near the shell aperture for shell secretion as the animal grows. Over the back of the drill and under the dorsal anterior portion of the shell and mantle there lies a large conspicuous chamber, the mantle cavity, which shelters a number of vital organs and openings. On the left side the mantle projects as a specialized tube into the siphonal canal of the shell. All inner surfaces of this tube and mantle cavity bear cilia which propel water by way of this tube into the mantle cavity and eject it from the right side of the mantle cavity. As water enters this chamber it first strikes a specialized sensory surface, the osphradium. This, according to the convincing studies of Yonge (1947), estimates the quantity of sediment carried in the water. Such a function appears of extreme importance to the drill which lives directly on the substratum where the danger of fouling or blocking of the mantle cavity is a constant one. It is supposed that in heavy suspensions of sediment the drill closes within its shell, although the reactions of Urosalpinx under these circumstances have not been described. That some drills do survive in densely turbid water has been demonstrated in portions of Delaware Bay where dense populations of Urosalpinx live in water so roiled much of the time that a Secchi disc disappears at 0.1 to 0.2 meter (T.C. Nelson, pers.com.

Immediately adjacent to the osphradium and extending the full length of the mantle cavity lies a large double-comb shaped gill. Probably much of external respiration is accomplished during the passage of sea water over this organ. Out of their native medium drills remain alive only so long as the mantle cavity is moist; by crawling about they accelerate the loss of water and hasten their destruction.

The anus in both sexes and the vagina in the female also open into the mantle cavity on the right side at the point where water is pumped to the outside. Another useful organ to the drill, a very large mucus gland, covers much of the dorsal half of the mantle cavity. It secretes copious quantities of a sticky fluid which entangles irritating sand and silt particles (Yonge, 1947) which in turn are removed from the mantle cavity by water currents on ciliary pathways. The presence within the mantle cavity of a well protected gill, an effective sediment testing organ in company with a highly efficient self cleansing mechanism, helps to explain the high degree of adaptability of this muricid to a wide range of habitats:

19

Internally the drill may be divided into two major body spaces which house the principal organ systems. The anterior space, or cephalic cavity, rests upon the heavy musculature of the foot, and contains the inverted resting proboscis, the central nerve ganglia, two pairs of salivary glands, the gland of Leiblein of unknown function, the esophagus with its attendant pharynx of Leiblein, and the large cephalic aorta. The posterior space, or visceral cavity, is located within the shell and contains the stomach, intestine, digestive glands, reproductive organs, heart, and kidney. According to tests by Sizer (1936) the flesh of drills maintains a salinity of 2.42 o/oo

Nervous System

The central nervous system of the drill consists of a number of ganglia which are concentrated in the head region in a mass the shape of a doughnut surrounding the esophagus and the large cephalic aorta. From this center large nerves radiate to all the principal organs of the body. The tentacles, front portions of the head and foot, siphon, osphradium, gill, and proboscis are especially heavily innervated. The proboscis alone possesses 7 pairs of distinct nerves which ramify into the principal structures of this active organ.

Urosalpinx possesses sensory organs which respond to at least four different kinds of stimulation: touch, sediment concentration, light, and smell-taste (chemical). Tactile organs are apparently present over the entire exterior surface of the soft parts of the drill. Although no experiments are reported, it may be assumed that the eye spots on the tentacles play a part in the response of the drill to light. Most external surfaces appear to be sensitive to strong chemical stimulation. Relatively dilute extracts of food (living or dead) may be detected principally by the anterior portions of the head and siphon. The proboscis probably plays but a small part, if any, in the initial location of prey, as it is lodged within the cephalic cavity and has no direct contact with the exterior when the false mouth is closed. The fully innervated tip of the everted proboscis is used in locating food close at hand and in selecting drilling points on the shell of its prey.

Circulatory System

The heart consists of a thin walled auricle and a strong muscular ventricle. A cephalic and a visceral aorta spring from the ventricle, the former to pass into the cephalic cavity and the latter to the organs in the visceral hump within the shell. The cephalic aorta runs forward through the central nervous system and there branches, sending one branch into the musculature of the foot and the other into the proboscis. These arteries play an important role in the movement and feeding of the drill. The artery passing into the foot transports blood which provides pressure to expand the foot as the animal emerges from its shell, and to evert the accessory proboscis. The artery running into the proboscis becomes relatively thin walled, muscular, elastic, and capable of considerable stretching commensurate with the flexibility of the proboscis. This artery funnels fluid to and creates pressure within the spaces in the odontophore which supports the radula. During drilling and feeding the radula is firmly supported upon the muscular odontophoral cushion made turgid by the action of numerous muscles contracting about the blood gorged odontophoral spaces. The proboscis as a whole is everted by combined muscular activity of the proboscis walls and by pressure resulting from muscular compression of the fluids in the cephalic cavity. Blood returns to the heart through a system of open spaces and short vessels among the organs.

Locomotory System

All shelled stages of the drill are capable of a slow creeping movement on the foot. The latter is a pale creamy yellow highly contractile muscular organ truncated in front and tapering behind, which scarcely extends beyond the broadest outlines of the shell. Federighi (1931c) observed that the drill moves by a smooth gliding motion and that the contact surface is covered with cilia. He was unable to detect pedal waves of muscular contraction during locomotion, and because the effective stroke of the cilia on the ventral surface of the foot is backward he supposed that locomotion is due to their activity. He noticed that at rest the snail is attached to the substratum by means of the posterior portion of the foot, and that when movement is initiated the anterior margin of the foot is extended forward and attached. Until the front part of the foot is in contact with the substratum no forward movement can occur. He observed in the laboratory that at 26.5°C the drill crept forward at an average rate of 2.6 to 2.8 cm./min. and did not creep backward. More recent observations suggest that drills may back into the bottom when burying for the winter (Carriker, 1954). Federighi states that adhesion depends entirely on the secretion of mucus as shown by the absence of areas of concavity which are necessary if suction plays any part in adhesion. The writer doubts, because of the unusual tenacity with which drills adhere to firm surfaces and because adhesion by muscular action does not necessarily produce obvious areas of concavity, that mucus is the sole agent of attachment in this case.

Galtsoff et al. (1937), also in the laboratory, and presumably at summer temperatures, noticed that the drill may move either on a horizontal or on a vertical surface at the rate of 2.5 cm./min., but point out that temperature and salinity of the water, character of the substratum, light intensity, and water currents may exercise an influence on the activities of the drill, hence its movements are necessarily variable.

21

Cole (1942) using British Urosalpinx, conducted laboratory studies at Conway on the rate of movement of the drill in a wooden trough filled with sea water. He found that individual drills varied considerably in the degree and rate of movement. The maximum rate (observed only once) was 3 cm./min. at 14.4°C. The maximum rate in the majority of experiments ranged from 1.17 to 1.75 cm./min. at all temperatures from 13 to 23°C.

Drilling and Feeding Organs

The oyster drill and similar predatory gastropods possess a feeding mechanism which exhibits a singular specialization of the muscular, nervous, and vascular systems in the proboscis and accessory proboscis in the direction of unusual adaptation to predation of specific prey. The effectiveness of this mechanism is amply attested by the wholesale depredations of young oysters on commercial grounds within its range.

The mouth parts of the drill are contained, not in the head region as in the majority of bilateral animals, but within the distal end of a long muscular tube, the proboscis, which consists in part of a modified extension of the ectoderm of the cephalic region (see plates 3, 4, and 8, in Carriker, 1943). The walls of the proboscis consist of an outer mucous secreting epithelium, and four thin layers of muscle, closely interwoven to form a tough pliant organ. This tube is capable of movement in all planes. In retraction the base of the proboscis is drawn in first and the tip follows last. The outer tip of the proboscis bears a heavily in nervated tactile rim, inside of which lies the true mouth. This opens into the buccal mass, a complicated muscular bulb which contains the tooth-studded radula. When fully everted the proboscis in a drill 35 mm. high measures 35 to 40 mm. in length, and the buccal mass in the tip measures 5 mm. in length, 2 mm. in width, and 1.5 mm. in height. In the posterior half of the buccal mass lies a fleshy tongue-like cushion, the odontophore, which supports the radula. A buccal cavity lies over the radula, and connects the mouth with the opening to the esophagus above and behind the radula. The latter consists of a uniform cylindrical translucent tube which passes backward from the buccal mass through the proboscis and the cephalic cavities to the stomach. The radula consists of a long narrow chitinous ribbon armed with three longitudinal rows of sharp hard teeth which point backward. Much of the radula is housed in a blind tube behind the buccal mass in the proboscis cavity and gradually grows forward out of this over the odontophoral cushion. The teeth are formed in this tube, and as the radula grows anteriorly the outer worn teeth are lost, probably swallowed, and replaced by new teeth. Chemical tests with acids show that silica is not present in the teeth in sufficient quantity to preserve their form in boiling concentrated sulphuric acid; rigidity of the teeth, as in other snails, probably results from impregnation with other inorganic compounds.

22

A striking complex of muscular bands controls the movements of the radula in drilling. Circular and oblique bands maintain the general shape of the odontophore and the buccal mass, and radial bands suspend the buccal mass in position within the proboscis. Short muscular bands passing forward from the odontophore to the tip of the proboscis, and very long bands extending backward through the proboscis cavity into the cephalic region, function coordinatedly in alternately drawing the radula forward during each rasping stroke and retracting it during the resting stroke. Likewise the radula is partly rotated on its long axis to either side during drilling to effect a round smooth hole. In addition to the tongue-like rasping motions of the radula, the radula itself moves back and forth independently over the odontophoral cusion at each stroke.

Two pairs of salivary glands are present in the forward confines of the cephalic cavity. One pair empties its secretions into the dorsal part of the buccal cavity, and the other into the true mouth. It is suggested that these secretions function principally in lubrication, although biochemical tests may disclose the presence of enzymes.

Drilling in Urosalpinx appears to be aided by the softening action of the secretions of the accessory proboscis which is located in the foot. This gland was first discovered by Fretter (1941) in the two British drills, Nucella and Ocenebra. The writer later discovered it in U. cinerea (1943). A minute constricted opening leads into the shallow chamber which encloses the excessively creased accessory proboscis. When everted in the living animal, this gland takes the form of a translucent white rounded projection slightly larger in diameter than that of the cephalic proboscis on the same drill, and with a height equal to the diameter. At first Fretter (1941) believed that this gland was concerned with the feeding process, but in a later paper (1946) she concludes that it is a sucker which is used to maintain a steady purchase on prey during drilling. Her conclusions are based on a histological study of the gland in the two British drills and in U. cinerea, and on two observations with Nucella in which this gland was seen to grip the shell of a mussel immediately below the spot at which the proboscis was at work. She notes that the surface of the gland is covered by a very tall epithelium composed of gland cells alternating with densely ciliated cells bearing short cilia. The secretion from these gland cells is exuded as a dense sticky substance which responds only slightly to stains specific for mucus. She also observed that in newly hatched drills the accessory proboscis is relatively very large, possessing a diameter equal to nearly 1/3 of the width of the foot. Fretter writes further that experiments (for which no details are given) with this gland show no solvent effect on the shell of other mollusks. However, studies which are being continued by the writer and are summarized in the following paragraphs, strongly suggest that the accessory probosics functions principally in the secretion of a substance which softens the shell preparatory to rasping.

23

Tarr (1885) in a superficial description of the structure of the proboscis gave a brief preliminary account of some of the aspects of the mechanical phase of drilling. By the use of a perforate cupped oyster valve containing a living shucked oyster and sandwiched between two microscopic slides immersed in a finger bowl of sea water with a few hungry drills (Prytherch's technic, unpub.), it has been possible more recently to observe the precise drilling behavior of U. cinerea under the binocular microscope (Carriker, 1943; and later, unpub.). When placed on living oysters some drills select the drilling point quickly and others search for some time before making the choice. During the search the proboscis is extended and its tip, undulating with minute wave-like movements, is passed slowly over the substratum. Using Prytherch's technic the writer soon discovered that drilling frequently is performed at the junction of the shell and one of the glass slides. The proboscis tip is extended to this site apparently attracted by chemical stimulation from the living oyster within, and since the glass slides partially bar the entrance of the snail shell, the proboscis and drilling are clearly visible. As drilling progresses it is soon evident that the radula is only very slightly effective in rasping through the calcareous layers of oyster shell, and that penetration is made possible through the activity of the accessory proboscis. After the drill has rasped the drilling site free of incrustations, periostracum, and soft shell material, it withdraws the proboscis and creeps forward until the accessory proboscis comes to lie immediately over the drilling site. The foot during this time adheres very strongly to the prey (this may account for Fretter's interpretation of the use of the gland); the accessory proboscis billows outward and completely fills the hole, remaining in this position for several minutes. During this time no noticeable movement of any part of the snail is evident, and the ventral surfaces of the foot remain tightly applied to the shell surrounding the hole, so that a watertight connection seems to be maintained. After a time the accessory proboscis is gradually withdrawn, the anterior part of the foot is backed away, and the proboscis is protracted. It soon locates the drilling site and continues rasping. This alternate process is repeated until the shell is perforated. Rasping intervals vary from 2 to 15 minutes and the alternate periods of softening, one to 47 minutes. Some unidentified chemical secreted apparently by the accessory proboscis appears to soften the shell material. This is suggested by the fact that after each softening period the radula removes microscopic flakes of the shell material which during the latter part of the previous rasping interval did not respond to rasping. It is possible, as suggested by T. C. Nelson (pers. com.) that secretions of the accessory proboscis acting on the conchiolin matrix of shell, free crystals of calcite, calcite ostracum, and chalky deposits in the shell (also see Galtsoff, 1954). Bits of shell material removed by the radula are carried back into the buccal cavity where suction from the esophagus draws them off the teeth and passes them into the stomach. The translucency of the proboscis permits observation of these functions. The frequency of the rasping strokes in an adult drill at 25°C was about 60 per minute.

24

Drilling, then, involves both a mechanical and probably a chemical phase in this gastropod, contrary to Korringa s (1952) suggestion that it is entirely mechanical (made without citation of the writer's 1943 paper). Reference should be made here to reports on the drilling mechanism of the boring gastropods of the family Naticidae (Jensen, 1951; Turner, 1953) which indicate that drilling is entirely mechanical; however, these investigations are not conclusive and should be extended.

The feeding process in the oyster drill consists of mechanical rasping of the softer flesh of the prey. The proboscis is extended through the newly drilled hole and the radula tears away bits of flesh with its sharp backward pointed teeth. Whereas during drilling the maximum stress is placed on pressing the radula against the substratum, in feeding it is directed to tearing off bits of flesh during the retractor stroke. The radula is ineffective in rasping tissues like the adductor muscle of adult oysters until they have undergone partial autolysis. The teeth thus do little more than rasp free soft tissues, incrustations, and softened shell material. Flesh caught on the radular teeth and transported into the buccal cavity is neatly removed by esophageal suction and then carried by ciliary and peristaltic activity to the stomach at an average rate of about 2 mm./sec. at 28°C. Loose food materials such as mucus and oyster ova are ingested mostly by means of the sucking movements of the buccal cavity and the esophagus while the radula remains stationary. Firm flesh is never "sucked" out of the oyster as was commonly reported in the earlier literature. Objectionable particles which pass as far as the esophagus are promptly regurgitated by a reversal of the movements of these organs. Rasping is a slow process, and since there is no crop in the digestive system and the tract is relatively small, in keeping with the carnivorous habit, the snail appears to be able to digest food and shell material at the rate at which it accumulates in the stomach.

Excretory System

The carnivorous habit of the oyster drill undoubtedly produces a plentiful supply of nitrogenous metabolic waste. Fretter (1946) in a brief though detailed account describes an accessory, and possibly the principal, excretory organ in U. cinerea. This is the anal gland, a brown or blackish tissue embedded in the wall across the upper posterior portion of the mantle cavity and underlying the rectum. In adults the gland is composed of a much branched system of blind tubules which coalesce and empty by way of a short duct into the rectum immediately behind the anus. The anal gland consists of only one type of ciliated cell whose cytoplasm becomes filled with brown spherical concretions. At what appears to be the beginning of a cycle these may be scattered irregularly, and later clump into a few larger masses in vacuoles. These concentrations, and sometimes whole cells, are expelled into the lumen of the gland and are directed toward the outside via the anus by languidly beating cilia. By use of trypan blue and soluble and insoluble iron saccharate Fretter demonstrated that the anal gland functions as a

25

kidney in abstracting excretory matter from the blood and in concentrating it in masses which can be passed readily from the body. She also showed that this gland is present, though in less developed form, in recently hatched drills. The digestive gland, some of the cells of which in some gastropods perform an excretory function, in Urosalpinx may function in this manner to a limited degree, since Fretter notes that in the drill the cells of the digestive gland appear to be extremely simple in structure.

Reproductive System

The sexes in the drill are separate and according to Federighi (1931b), who studied 1,121 drills in the Woods Hole area, they occur in about equal numbers, but females are slightly wider and attain a greater height than males. Stauber (1943), however, found that a greater proportion of large drills collected in Delaware Bay were females, although he did dissect some males over 30 mm in height which suggested to him that a certain amount of protandry may occur. Table 1 gives a representative sample of Stauber's measurements relating height and sex in Urosalpinx. Because of difficulty in handling, no drills smaller than 17.6 mm. in height were included. The doubtful column represents drills too badly crushed after opening to make diagnosis of sex satisfactory. Stauber notes that although the mean height of the males and females does not differ significantly in Table 1, numerous repetitions of such data give a similar distribution of height. Stauber's measurements were made on drills trapped in April; whether drill trapping is selective for either sex is not reported, but should be kept in mind here.

The large curved penis in the male, though not easily seen, the yellow to orange colored female gonad, and the whitish male gonad (Federighi, 1931c; Cole, 1942) afford reliable characters by which the sexes may be distinguished.

There are no studies available on the reproductive organs of U. cinerea. Fretter (1941) presents an admirable treatment of this subject in the closely related English drills Ocenebra and Nucella; and since in general the structure and functioning of this system in these three closely related drills may be similar, it is instructive to briefly review Fretter's studies here.

The male reproductive system includes a testis where sperm are formed; a ciliated duct, the vas deferens, which carries sperm to the large prostate gland where seminal fluids are added; from the prostate gland sperm are transported along another ciliated duct to the penis which transfers them to the vagina of the female during copulation. Andrews and McHugh (pers. com.) report that examination of drills by cracking off the shell revealed that Urosalpinx as small as 9 mm. in height possess a well developed penis.

26

TABLE 1. The Relationship of Height and Sex
in Urosalpinx cinerea from Delaware Bay, New
Jersey (Modified from Stauber, 1943)

Height of Shell	Frequency		
in mm.	Males	Females	Doubtful
17.5 – 18.5	1		
18.6 19.5	2		
19.6 20.5	1	2	2
20.6 21.5	5	4	6
21.6 22.5	4	14	3
22.6 23.5	5	6	1
23.6 24.5	3	21	3
24.6 25.5	4	18	2
25.6 26.5	1	10	
26.6 27.5		11	1
27.6 28.5		$\frac{5}{3}$	
28.6 29.5			1
29.6 – 30.5		-	
Total number of drills	26	96	19

27

In the female reproductive system ova are produced in an ovary and from thence pass down an oviduct into an albumen gland where they are surrounded by an albuminous secretion. Since the eggs are enclosed within a common mass of albumen, embryonic cannibalism is made possible during subsequent stages of development. Fertilization of the ova probably occurs in the albumen gland. Next the mass of eggs suspended in albumen is moved by muscular and ciliary activity into a large capsule gland which secretes a wall of mixed proteins and mucoid materials about them. At one end of the capsule a plug of mucus is moulded into position by a special part of the gland. The capsule gland connects with the outside by way of a short tube which terminates in the vaginal opening. During copulation seminal fluid is received within the vagina and from thence at a later time sperm move across the capsule gland to a seminal receptacle where they are stored.

The vagina opens to the exterior on the right anterior side of the mantle cavity, ventral to the anus. The egg capsule, roughly formed by the capsule gland, is forced out of the vagina and is carried along a temporary groove on the right side of the foot to the pedal egg capsule pouch in the sole of the foot. Within this it is moulded into its final shape, its walls are hardened, and adhesion to the sub-stratum is effected.

Although Fretter reports that copulation is of frequent occurrence in both English drills, it has never been recorded for U. cinerea. A single labor-atory observation in Eupleura caudata, close relative of Urosalpinx, has been described by Stauber (1943) and is worth repeating because of its possible similar-ity to that in Urosalpinx. Preparatory to copulation a small male Eupleura mounted a large female and assumed a position on the right anterior side of the shell. During copulation the front mid longitudinal third of the male's foot was depressed into a groove and the penis was extended over this and around the rim of the shell of the female into the vaginal opening. After numerous interruptions the male was repeat-edly successful in relocating the female and persisted in copulation. Under the conditions of these observations copulation continued intermittently for 21 days. A study of the gonads and behavior of representative stages in the life cycle of Urosalpinx isolated in the egg case stage should contribute materially to our knowledge of reproduction in this drill

At first Cole (1941) considered that sex reversal might occur in U. cinerea, but further work by him (1942) demonstrated to his satisfaction that there is no evidence to support this hypothesis. Twenty drills caught in the act of spawning were isolated for the summer and showed no sexual change by fall. In addition a careful search among thousands of drills disclosed no individuals with characters inter-mediate between those of male and female. In confirmation of these studies drills should be isolated over a longer period of time and histological studies of the gonads of sample drills should be made periodically.

28

Stauber (1943) reports observations which suggest that sperm of Uro-salpinx may remain viable in the female for extended periods of time. He isolated an adult female collected in Delaware Bay from April 9 to October 21. During this interval the snail oviposited 96 egg cases from which active young drills were subsequently hatched in the laboratory. It is not known whether the drill copulated in the early spring before capture or during the previous season, but because of the low temperature of the water in April it is more likely that copulation took place during the previous season.

Ova

Ova when first oviposited in the egg case are spherical in shape, average 0.36 mm. in diameter, and are yellow to orange in color. Reports on ova production in the drill in different geographical areas are tabulated in Table 2. Stauber (1943) in the course of numerous careful measurements disclosed that, in general, larger drills oviposit larger egg cases which in turn contain more ova than do capsules of small drills: the average number of ova laid per egg case by a 16.5 mm. drill was 4.7 and by a 29.6 mm. drill, 11.5. In addition the number of ova per egg case seems to be influenced by unknown functional factors. Stauber (1943) and Haskin (1935) independently describe instances where no ova were found

TABLE 2. Tabulation of the Egg Production in Urosalpinx cinerea in Different Geographic Regions as Reported by Numerous Workers

Number of Ova			
Average per egg case	Range per egg case	Region	Source
8.5		Eastern Canada	Adams, 1947
10	5 - 17	Woods Hole, Mass.	Pope, 1910-11
8	4 - 16	Barnegat Bay, N.J.	T.C.Nelson, 1922
8	0 - 20	Barnegat Bay, N.J.	Haskin, 1935
8.1	0 - ?	Delaware Bay, N.J.	Stauber, 1943
11	6 - 20	Chesapeake Bay, Va.	Brooks, 1879 (1880)
8.8	3 - 22	Hampton Roads, Va.	Federighi, 1931c
11.7	1 - 29	England	Cole, 1942

in egg cases; such abberations seem most likely to occur during the initial attempts at oviposition in the life cycle of the drill. Haskin's (1935) data in Table 2 is based on a collection of 1, 297 egg cases in Cedar Creek, Barnegat Bay, New Jersey, between June 20 and July 4, 1935. Cole's (1942) figure of 11.7 for the average number of eggs per case was computed from a collection of 1, 423 egg cases from different localities and is similar to Brook's (1879-1880) figure. Stauber's data would suggest that these high figures may have resulted from collections of egg cases oviposited predominantly by larger females. Cole reports that there is little variation in the number of eggs per case in eggs collected from one parent; this is probably for egg cases oviposited at one stage in the life cycle of the drill. The overall average of ova per case in Table 2 is 9.3. The variations in the reported regional averages are not excessive and may express differences in age of the ovipositing females except in the English Urosalpinx which because of its larger size may oviposit more ova per capsule than its American relatives.

Egg Capsule

The egg cases of Urosalpinx are tough, leathery, urn shaped capsules which occur in clusters tightly affixed to firm substrate by means of short slender stalks whose bases unite with those of neighboring egg cases. A round lid or operculum is located on the middle free end of the case and is described by Pope (1910-11) as similar to a door which after the emergence of the first drill hangs as if by a hinge. The wall of the egg case consists of three layers: an outer tough one is detachable by mechanical means; a middle layer, also tough, is transparent so that by removal of the outer membrane the embryology of the drill may be observed (Haskin, 1935); the innermost lay, first described by Haskin (1935), is a delicate membrane which completely encloses the eggs and is present throughout the development of the embryos, although it becomes invisible in the later stages of development. Haskin concludes that the innermost membrane may play a major role in the permeability of the egg capsule. Egg cases are clear bluish white when first laid but gradually change in color through yellow to a deep yellowish brown at the time of hatching. Ova are suspended within the egg case in a soft transparent jelly-like medium which serves as a source of food and a buffer against mechanical shock

Egg case membranes are composed of relatively insoluble protein, and are permeable to the constituents of sea water as well as to a variety of foreign inorganic solutions, organic salts, and dyes, some of which may be toxic to the developing drill. The cases seem to function in protecting the larvae from mechanical injury and from predatory organisms but not from ionic changes in the environment. The salinity of the contents of the egg case is much lower than that of sea water, and may be related to the low salinity of the prehatched drills (Sizer, 1936; Galtsoff et al., 1937).

30

Pope (1910-11) records the size of the drill egg case in the Woods Hole area as averaging 7 mm. long, 4.5 mm. wide, 1 mm. thick, with an operculum 1 mm. in diameter. Stauber (1943) observed that as drills grow larger they lay bigger egg cases. For example in Delaware Bay, New Jersey, a drill 16.5 mm. in height deposited egg cases averaging 3.8 mm. in length, and a 29.6 mm. drill, egg cases 8.4 mm. long. Drills under 20 mm. in height rarely deposit egg cases over 6 mm. long and those over 25 mm. seldom lay cases smaller than 6 mm

The number of egg cases produced by Urosalpinx in a number of different regions within its geographic range is reported in Table 3. Egg production is not necessarily uniform over a period of years, as Stauber (1943) observed that the average number of egg cases oviposited by caged drills in Delaware Bay in 1941 was considerably lower than that in 1940. A single female does not always deposit all her egg cases at one time, but may lay a number of clutches during the breeding season (T. C. Nelson, 1922; Engle, 1940; Cole, 1942); and the number of egg cases laid in a cluster at one time is quite variable. Pope (1910-11) noted that egg laying will often extend for a period of several days, and he found a minimum of four and a maximum of 150 egg cases in clusters. A number of females may oviposit together and this probably accounts for Pope's maximal figure. In detached clusters laid by isolated females the range in the number of egg cases varied from 4 to 22. Cole (1942) observed that capsules are deposited at the rate of 3 or 4 per day.

The data tabulated in Table 3 is not entirely comparable. T. C. Nelson (1922) observed oviposition for only about a month, which probably did not include the total egg laying period; Galtsoff (Galtsoff et al., 1937), according to Stauber (pers. com.) began his observations in 1935 after oviposition had started and terminated them in June, 1936, before oviposition had ceased; and Adams (1947) has recorded only a limited quantity of data. A possible source of error is introduced by the lack of information on the relation of the number of egg cases deposited per season to the age of the drill, although it is likely that up to a point the rate of oviposition may accelerate with age. A serious source of error in all these estimates of oviposition is found in the lack of information on the proportion of drills under observation which were females. Further, it has not yet been demonstrated by controlled experiments that confining drills in cages does not alter total seasonal oviposition. At best these preliminary data tentatively indicate that Urosalpinx deposits an average of approximately 45 egg cases per season, starting with a minimum of zero in immature females and reaching a possible maximum of 96 cases per season in older mature females.

TABLE 3. The Number of Egg Cases Oviposited by Urosalpinx cinerea in different Geographic Regions as Reported by Several Investigators

Number of Egg Cases
Deposited per Drill

Per Day	Average per Season	Range per Season	How Data Obtained	Region	Source
	–	--67	.	Eastern Canada	Adams, 1947
	50.6		Drills isolated in field cages, Apr. 13 to May 16.	Little Egg Harbor, N. J.	T. C. Nelson, 1922
		--50	Isolated females.	Delaware Bay, N. J.	Galtsoff et al., 1937
	72		Field, estimate for females over 21 mm., 1940.	Delaware Bay, N. J.	Stauber, 1943
	57.6	30–96	Drills isolated in field cages, 1940.	Delaware Bay, N. J.	Stauber, 1943
	34		Drills isolated in field cages, 1941.	Delaware Bay, N. J.	Stauber, 1943
3.9	28		Isolated drills.	Hampton Roads, Va.	Federighi, 1931c
3–4	25	20–35	Isolated drills.	England	Cole, 1942

ECOLOGICAL LIFE HISTORY
Duration and Intensity of Oviposition

Information on the duration of the spawning period of the oyster drill in native waters in different latitudes is abstracted in Table 4. The noticeable variations in the onset of spawning in one region as well as along the entire eastern coast of North America and in England undoubtedly reflects not only annual differences in water temperatures in the spring but incomplete information as well.

For example, Gibb's date represents only a single observation, and Bumpus and Federighi, the latter for Beaufort, give no termination dates for the spawning period in these states. Engle's (1940) date for initiation of spawning in Long Island Sound is almost a month later than Loosanoff's date for the same sound; it is possible that Engle did not observe the earlier spawnings, or that, since spawning in warmer inshore waters occurs at an earlier date than in the offshore colder waters, the two sets of observations were made in waters differing considerably in their thermal characteristics. However, Loosanoff made his records over a period of several years and thus probably more closely reflects the overall spawning picture in the area. The spawning date in the spring in Delaware Bay given by Galtsoff et al. and by Stauber differ by almost a month; Stauber (pers. com.) notes that the date of Galtsoff et al. is not representative for Delaware Bay. Again because of Stauber's long range studies in this bay it is probable that his observations more accurately reflect the conditions there.

In Cape Cod waters Galtsoff et al. (1937) found that the bulk of the spawning occurred in the early part of the summer, and that a second smaller spawning occurred in late September. Cole (1942) made a similar observation in England where he found that the bulk of spawning took place during May and June after which it declined until late August and September when a second much smaller spawning took place. The same phenomenon is reported by Stauber (1943) for Delaware Bay where his rather complete information shows that over a period of years spawning began in May, reached a peak in June, and ceased almost entirely in August; and a second less intense wave appeared in September and this ceased in October or November, depending on the temperature of the water. Stauber found that the second wave of spawning was performed chiefly by young drills maturing in the late summer. Galtsoff et al. (1937) observed a more or less continuous spawning over a period of 7-1/2 months in Chincoteague Bay, Virginia, with a climax during June and July; and Federighi (1931c) noticed that spawning continued throughout the summer in Hampton Roads, Virginia, and in Beaufort, North Carolina, and gradually decreased in intensity in the fall. Neither mentioned a second peak of spawning in the fall. The occurrence of late spawnings in at least three different regions suggests that maturing young drills may oviposit in the late fall throughout much of their northern range.

33

TABLE 4. Duration of the Spawning Period of Urosalpinx cinerea in Native Waters in Different Latitudes within its Geographic Range as Reported by Various Investigators

Spawning Period	Period of Maximum Spawning	Region	Source
Late Apr. .... Sept.	May–June	England	Cole, 1942
June 1 .... Sept.		Eastern Canada	Adams, 1947
June .... Sept.	July-Aug.	Cape Cod, Mass.	Galtsoff et al., 1937
May 21 .... ?		Woods Hole, Mass.	Bumpus, 1898
June 1 .... late Aug.		Woods Hole, Mass.	Pope, 1910-11
May 11 .... ?		Narragansett Bay, R. I.	Gibbs (Stauber), 1943
June 15 .... Sept. 15		Long Island Sound, N. Y.	Engle, 1940
May 20-29 ...... late Oct.		Long Island Sound, N. Y.	Loosanoff, 1953
Early Apr. ..... late Nov.		Little Egg Harbor, N. Y.	T. C. Nelson, 1922
Apr. .... late Nov.		Delaware Bay, N. J.	J. R. Nelson, 1931
Early Apr. ..... late Nov.		Delaware Bay, N. J.	Galtsoff et al., 1937
Apr. 26-May 16...Oct.-Nov.	May-July	Delaware Bay, N. J.	Stauber, 1943
June .... Jan.	May-July	Chincoteague, Va.	Galtsoff et al., 1937
May 20 .... Oct. 1		Hampton Roads, Va.	Federighi, 1931c
Mar. 31 .... ?		Beaufort, N. C.	Federighi, 1931c
? .... ?	Mar.-May	North Carolina	Galtsoff et al., 1937

Late warm periods in the fall have been the apparent cause of a number of unusually late spawning records. T. C. Nelson (pers. com.) once caught a drill ovipositing on December 26 in New Jersey during a very warm period. The egg cases lacked opercula and cleavage of the eggs was abnormal. Andrews (pers. com.) recently found a few egg cases with live embryos in Mobjack Bay, Virginia, in February. And Loosanoff and Davis (pers. com.) have found recently oviposited egg cases in Milford Harbor as late as November 19, also probably as a result of warm weather.

The periods of maximum spawning in different areas given in Table 4 are probably not comparable because no quantitative standard is given by any of these investigators for "maximum" spawning. Nonetheless these periods and the dates at which spawning is said to commence in various regions both appear, though nebulously and with numerous exceptions, to occur earlier in the year in the southern latitudes than in the more northern waters.

Behavior of the Drill during Oviposition

Female oyster drills generally affix their egg cases to the sides, under-surfaces, and crevices of mollusk shells, cement blocks, tin cans, rocks, stakes, piling (Galtsoff et al., 1937), and any other hard available surfaces which may be only partially submerged, as in the lower intertidal zone, or completely submerged, as on subtidal grounds. Egg case clusters may also be found abundantly under rocks (Pope, 1910-11).

In addition to seeking hard surfaces for oviposition, drills generally select sites which project somewhat above the surface of the bottom and which offer niches free from siltation and possible burial and suffocation (Federighi, 1931c; Cole, 1942; Stauber, 1943). In the laboratory egg cases are most frequently deposited on the vertical sides of aquaria if no clusters of oysters or similar objects are present on the bottom

However the presence of suitable food material may further influence the selection of the spawning site. Federighi (1931c) and the writer observed that drills in laboratory tanks in which living oysters are present in almost all cases crawl onto the living oysters to lay their egg cases in preference to the sides of the tank. Sizer (1936) reports that in drill trapping experiments in Delaware Bay he found the upper valve of a living oyster is preferred to the surface of an empty shell for oviposition. Stauber (1943), who continued these studies, found in June, 1937, while dredging over bottom covered mostly with shells that egg cases were predominantly attached to living oysters: of 301 shells examined only one possessed egg cases, and of the five large oysters in the same catch, two held egg cases.

35

Similarly he found that during two successive seasons oyster baited traps attracted more oviposition than shell baited traps. In 1939 during the major spawning period 250 oyster baited traps caught 1.03 drills and 0.03 cluster of egg cases, and 50 shell baited traps captured only 0.45 drill and 0.01 cluster of egg cases per trap per week.

As many as a dozen female drills may aggregate to spawn on one location (Pope, 1910-11, Massachusetts; Cole, 1942, England; the writer, in aquaria in New Jersey and in North Carolina) Cole does not think this represents a social behavior but merely the occupation of a nearby available and satisfactory spawning site. This, however, does not seem to be the fundamental explanation; further research should be performed on this interesting and significant behavior. Preparatory to deposition of her egg cases, the drill by the use of her radula carefully cleans the surface to which the egg cases are to be attached. If disturbed while spawning she may creep away, but in many cases returns in a few days to the original site to resume spawning. If undisturbed she has been observed to spawn continuously for as long as 7 days (Galtsoff et al., 1937; Federighi, 1931c). Federighi noticed that sudden drops in temperature or lifting the drill from the substratum stops spawning.

In an effort to study the relation between the quality of food which drills eat and fertility, Galtsoff et al. (1937) placed lots of 10 drills each in tanks supplied with running sea water containing a variety of food animals. It is not stated how the sex of these drills was determined or how many were immature females. The authors conclude because they obtained considerable variation in the number of egg cases deposited that fertility in drills is correlated with the quantity and quality of food. On the basis of Stauber's (1943) data on the relation of size and maturity of drills to oviposition, and because there is no assurance that all 10 drills in each experiment were females (many may have been males), some doubt is cast on these results. Nonetheless, as Stauber states, the quantity and quality of food probably do influence oviposition. Although Pope (1910-11) noticed no cessation of feeding of a number of drills during the entire spawning season from June 1 to August 1, it is true that females do not feed during the actual process of oviposition (Federighi, 1931c; the writer, unpub.), and probably feed more rapidly after spawning.

Egg Case Stages

After the egg case is affixed to the substratum drills exhibit no concern for the young. Egg cases are abandoned and the developing young, which pass all the larval stages within the egg case, care for themselves. The absence of a planktonic stage, though limiting rapid dissemination, has not appreciably interfered with the success of this animal. If anything, the protective confines of the egg case support a high rate of survival of the young.

Although cloistered, not all the fertilized ova achieve full development. In a general account of the early embryology of U. cinerea Brooks (1879-1880) was the first to show that abnormal eggs are frequently found in egg capsules, and that occasionally an advanced embryo breaks up and these swimming fragments are consumed by other embryos. Pope (1910-11), Federighi (1931c), Haskin (1935), and Cole (1942) have confirmed these observations.

Federighi (1941c) reported that 28 egg cases containing an average of 8.8 ova per case gave rise to an average of 5.1 young drills per case, suggesting a mortality of about 42% in the waters of mid eastern United States. Haskin (1935) working on drills in New Jersey found that approximately half of the eggs laid in egg capsules failed to mature. He suggests that the existence of some atypical sperm in the gonads of males may be associated with the development of abnormal embryos and this has prevented unrestricted cannibalism. Brooks earlier (1879 - 1880) indicated that he thought this method of feeding is an accidental and exceptional one. To the writer's knowledge no one has described atypical sperm in Urosalpinx. Such sperm should be looked for and, if present, their relationship to prehatching mortality determined. Cole (1942) in English waters observed an average mortality during incubation of 1.73 embryos per case, or 13.9%. Of a total of 1,423 egg cases collected in different localities, 823 cases contained shelled veligers about to hatch which averaged 10.74 per case. Cole's mortality figure of 13.9% is noticeably lower than the figures of Haskin and Federighi, and since it is based on extensive field observations is perhaps the more realistic; however, it is also possible that in addition a regional or a racial difference is being expressed here.

Incubation Period and Hatching Behavior

Data on the duration of the incubation period of the egg case stages of the drill reported for a number of geographic regions, both in the field and in the laboratory, are summarized in Table 5. Federighi (1931c) determined the incubation period in the laboratory in Hampton Roads by isolating 11 different groups of freshly laid egg cases at different intervals between May and August. Haskin (1935, quoted by Galtsoff et al., 1937, without citation) in Cedar Creek, New Jersey, in the field at a carefully recorded temperature range, found that the first drills may hatch from different clusters over a period of 18 to 31 days, and that normal late stages could still be found in some egg cases in the same cluster as late as 46 to 53 days after oviposition. Stauber's (1943) data are based on the first appearance of egg cases and of small drills 2-3 mm. in height on the bait of drill traps in large scale field operations, and thus the duration of the incubation period which he records is understandably longer than that of Haskin (1935) even though Haskin's observations were carried out at a slightly lower temperature range. Stauber noted, as did Pope (1910-11), that in the fall at

37

TABLE 5. Duration of the Incubation Period of the Egg Case Stages of Urosalpinx cinerea as Reported by Different Investigators for a
Variety of Geographic Regions

Duration of Incubation Period, Days	Temp. °C	Site	Region	Source
49-56	13.5-19	field	British waters	Cole, 1942
27-32	22.6	lab	British waters	Cole, 1942
44-50	18.3	lab	British waters	Cole, 1942
45-56		field	Eastern Canada	Adams, 1947
35-49		field	Woods Hole, Mass.	Pope, 1910-11
35	?	lab	Woods Hole, Mass.	Pope, 1910-11
about 35	?	field	Barnegat Bay, N. J.	T. C. Nelson, 1922
18-53	23.3-29.1	field	Barnegat Bay, N. J.	Haskin, 1935
45-78	15-25	field	Delaware Bay, N. J.	Stauber, 1943
36-44	18-32	lab	Hampton Roads, Va.	Federighi, 1931c
26-38		lab	York River, Va.	Newcombe, 1941-42

temperatures below 15°C development proceeds very slowly. Newcombe (1941-42) in an examination of egg cases from three different oyster grounds from July 7 to August 8 in the vicinity of York River, observed that a high proportion of the egg cases contained hatching drills throughout this period. The highest percentage of hatching drills, 62%, was taken on August 8.

Cole's (1942) controlled laboratory experiments demonstrate the marked relationship of temperature and the duration of the incubation period. The close similarity of the duration of the incubation period at a relatively constant temperature of 18.3 in the laboratory and at a fluctuating temperature range of 13.5-19°C in the field is equally striking. In addition his field data suggest that development in English waters proceeds at a lower temperature than in the waters of the middle eastern United States (compare with Haskin, 1935; Stauber, 1943; and Federighi, 1931c)

When considered from the standpoint of completeness of observations and relation of these to fairly typical ecological conditions, the work of Cole (1942), Haskin (1935), and Federighi (1931c) seems to most clearly describe the duration of the incubation period in these regions. More carefully controlled experiments on incubation in different geographic regions are suggested by these studies, many of which, though important in a preliminary way, are quite incomplete.

The prehatched shelled stages of the drill are known as protoconchs and the hatched stages as conchs. Pope (1910-11) described the emergence of proto conchs in considerable detail. Prior to hatching, the orifice of the egg capsule is closed securely by the thin operculum. The first protoconch to emerge pushes it outward and others follow. Then for a brief period the newly hatched conchs cling to the sides of the parent case. No one has observed whether the young drill cuts the operculum open with its radula or whether by hatching time the periphery of the operculum has been freed by some action such as bacterial activity. Proto-conchs can also effect their escape from the egg case by drilling. In the laboratory Pope watched a number of young drills cut small circular holes the diameter of a cambric needle and push their way out through this, leaving warty protuberances on the case at the site of escape. Pope believes this mode of escape occurs when protoconchs near the orifice either obstruct the passage or are not developed sufficiently to emerge. The interval for all drills to hatch from an egg capsule or cluster of capsules varies considerably because of the uneven development of the embryos, and may extend from four to 38 days (Pope, 1910-11).

The degree of winter survival of young drills hatching from the late summer and autumn wave of oviposition in northern waters has never been determined. Stauber (1943)states that it is quite doubtful that there is any

39

appreciable survival since he encountered so few of them in spring and early summer sampling in Delaware Bay. In addition the second spawning of the season is slight and probably does not materially augment the drill population.

Growth

Information on the rate of growth of U. cinerea is grossly incomplete particularly for drills in North American waters. Likewise little is known about the relationship between growth rate and such factors as kinds and quantity of food, sexual activity, temperature, salinity, turbulence, turbidity, and substratum.

Height of shell (more popularly referred to as "length" of shell), from the apex of the spire to the tip of the siphonal canal, has been widely used in reporting size data on drills. For young snails which grow relatively fast this dimension has proved useful. but for larger gastropods Stauber (1943) suggests that volume is probably a better measure, since wearing of the shell may actually indicate a decrease in shell height. In the case of large Urosalpinx a small increase in shell height is attended by an appreciable increase in volume.

The employment of shell size, either height or volume, as a criterion in the determination of drill age is unsatisfactory because young drills hatch continuously throughout the warmer months of the year and thus provide a wide range of sizes in each year class. Stauber (1943) in Delaware Bay repeatedly obtained drills as small as 2 mm. in late July and August, and those only 4 mm. in April which were probably winter survivors of the previous late summer ovipositions. T. C. Nelson (1922) collected Urosalpinx, presumably hatchlings of the previous summer, whose height ranged from 6 to 10.5 mm., average of 9 mm., in April in Little Egg Harbor, New Jersey. Considerable individual variation also occurs in growth rate probably not only in the presence of variable food and other ecologic conditions but because of individual genetic differences. Stauber reports that drills survived for 19 months in his laboratory without food; thus in the field they could survive for long periods under poor food conditions without an appreciable increase in height.

A little information is available on the rate of growth of Urosalpinx in America in the first year or two. In northern waters Pope (1910-11) and Stauber (1943) report that on emergence from the egg case the young conch varies in height from 1 to 1.5 mm. In southern waters Federighi (1931c) noted that newly hatched drills average 0.8 to 1 mm. in height. Pope observed that they double their size in 8 to 10 days in the laboratory. He assumed that since the smallest drills found in early June measured 12.5 mm., this size represents the growth of one year. J. R. Nelson (1931) writes that drills collected in

40

September and hatched the spring of that season measured 8 mm. in height, and accepts this as the approximate growth during the first summer in Delaware Bay. Adams (1947) in eastern Canada finds that drills grow rapidly and reach a height of 13 to 19 mm. during the first growing season. He does not describe the method by which this information was obtained.

Stauber (1943) was able to obtain some data on the growth rate of older drills caged with oysters in Delaware Bay from May 24 to August 8, 1939. During this interval, 21 drills ranging in height from 11-15 mm. (means 13.4+0.8) reached a height of 16-24 mm. (20.7+1.6); 13 drills survived. Another group of drills 21-30 mm. (25.7+1.6) in height attained a height of 23-30 mm. (26.3+1.4). The increase in height in the smaller drills was about 55% of the original height, and it was slight in the larger drills. In an unusual case a female isolated from April to October increased in shell height from 15.2 to 31.4 mm. The possible inhibiting effect of confinement and crowding on rate of growth is a factor which might well be considered in future growth studies.

The most complete data available on the growth rate of U. cinerea was obtained by Cole (1942) for drills in English waters. Cole's fine report presents a careful analysis of growth in males and females based upon growth marks and on size distribution curves. He noticed that these curves usually show a number of closely approximated peaks which when correlated with growth marks appear to approximate annual growth increments. Clear growth marks may frequently be seen on the tip of the shell bounding the siphonal canal and occur more closely spaced after the first few years of life. On large shells these occur at 1 to 3 mm. intervals. Since considerable overlapping of successive year groups probably occurs which seriously impedes or entirely prevents the fixing of the position of the peaks in the frequency curves, Cole made use wherever possible of the size distribution curves by the freehand method advocated by Buchanan--Wollaston and Hodgson (1929). His growth data represent measurements of some 1,700 drills which were collected principally by hand picking in the intertidal zone of the River Blackwater and the River Roach, Essex, during the warmer months of the year over a period of three years. All drills visible during low water of spring tides were included in each sample. In 1941 samples were dredged at monthly intervals in the River Blackwater.

This information is summarized in Table 6. Cole found no substantial number of males over 36 mm. or females over 39 mm. He confirms earlier reports that females grow more quickly than males and reach a larger size, and shows that in general U. cinerea reaches a greater average size in Britain than on the Atlantic coast of North America. In laboratory checks on growth rate he reared drills hatched in July on small oyster spat in a plunger jar at Conway. By the end of the first feeding period these drills reached a maximum height of 12 mm., the mode

41

TABLE 6. Growth Rates of Urosalpinx
cinerea in English Waters as Determined by Cole (1942) by Means of
Size Frequency Curves and Growth
Marks

Probable Age in Years, in July	Average Height, mm.	
	Males	Females
1st. summer	10	–
	10-18	20
2	18-23	25
3	23-25	27
4	25-27	29
5	27-29	31
6	29-31	33
7	31-32.5	35
8	32.5-34	36.5
9	34-35.5	38
10	35.5-36.5	39
11	37.5	40
12	38.5	41
13	39	42
14	–	43

of the group being approximately 10 mm. He suggests that they might be expected to reach a height of about 20 mm. by the following July, the end of their first year. He found that samples of drills dredged during the summer contain appreciable numbers of small drills which usually have a suggestion of a peak around this figure ... he was unable to collect adequate samples of the small drills because the majority passed through the rings of the oyster dredge used in sampling.

Hancock (pers. com.) confirms Cole's results and believes his method of analysis of growth rates is reliable. Hancock adds that in the planning and evaluation of such studies a number of factors should be considered: (1) young drills hatch throughout much of the summer, and since the largest growth incre ments are added in the first two years, there is probably a marked size variation among individuals of any given population throughout the season; (2) in a proportion of specimens a thickening of the lip of the shell takes place and growth appears to cease; (3) populations less than three miles distance from each other in the same river, and those from shallow and deep water in the same area of the river, are characterized by quite different average and maximum sizes; thus great care is necessary in sampling a limited area, and in the choice of methods of sampling which are employed.

From the incomplete information available on growth rates in Urosalpinx it may be tentatively suggested that in America the drill may reach a height of 8 to 19 mm. in the first summer (J. R. Nelson, 1931; Adams, 1947). The unusual growth of a single female drill in Delaware Bay from 15.2 to 31.4 mm. in one summer (Stauber, 1943), coupled with the possible maximum rate of growth during the first summer, indicates that in America the fastest growing females may achieve a height of 31 mm. in two years. This is in marked contrast to Cole's (1942) data which suggest that in England it takes about four or five years for a female drill to attain this height. It is difficult to comprehend that drills originating in America should exhibit such retarded growth rates in English waters; the available data is probably too inadequate to permit such comparisons.

As a quantitative controlled check on the growth studies of Cole and others, and to provide accurate data on growth rate, size and age at sexual maturity, maximum size, and longevity of Urosalpinx, it is urged that Urosalpinx be reared in the laboratory from the egg stage to senescence in isolated running sea water containers. This drill is easily cultured in the laboratory, and a number of permanent marine laboratories now have facilities in which it would be possible, if necessary, to maintain populations of drills for as long as 15 years.

Walter (1910) in a comprehensive study of the ratio of maximum shell aperture to shell height in 30,903 U. cinerea collected in four different localities in the Woods Hole area over a period from 1898 to 1908 observed that as drills in a given population grow larger this ratio diminishes regularly. He noted this trend in collections taken at biweekly intervals during the summer, and also in collections made successively over a period of years during the first week in August. In the course of 7 years of successive collecting he found that this ratio fluctuated regularly and reached the highest average in 1902. Walter implies that this probably is a reflection of the yearly increase in size of the majority of the drills in the population of a given locality and that the high average will repeat itself in future years as the cycle repeats itself. He believes that this gradual change in form as the drill grows larger is related to internal developmental changes occurring during the life history of each drill and is independent of environment

Growth rate, temperature, and latitude

Rate of growth in the oyster drill is strikingly influenced by temperature, and it is very unlikely that size increases occur during the colder months of the year (Stauber, 1943; Pope, 1910-11). The existence of mean size variations in drill populations in different geographic regions, which at first suggested that drills grow to a larger size in colder waters, led to a number of studies. Federighi (1931c) seeking to explain these geographic variations on the basis of environmental differences alone, was the first to suggest that drills might grow to a larger size in waters of lower temperatures since he (1931a) had noticed that drills in North Carolina grow to an average size of 15 mm. in height, and those in Virginia to 23 mm. in water colder by a mean temperature of about 4°C. Fraser (1930-31) from a study of 1,000 drills in Essex, England, obtained a mean height of 30 mm. Comparing his results with Federighi's he states that there does seem to be some temperature correlation, the means of approximately 11, 17, and 20°C corresponding to the mean heights of 30, 23, and 15 mm. in Essex, Virginia, and North Carolina respectively. He admits that this correlation may be only superficial, although there is no doubt that the drill grows to a much larger size in England than in most American waters. The giant drills averaging 44-50 mm. in height which grow on the eastern shore of Virginia (Baker, 1951) are probably a different subspecies and should be considered in a separate category. Not long after, Federighi (1931b) had opportunity to measure over a thousand drills from Woods Hole in water colder than that in Virginia and North Carolina and obtained a mean height of only 21 mm. He rightly points out that since salinity and possibly other ecological factors vary among these areas it is problematical what factors influence drill size.

44

These preliminary data are much too incomplete and collected too variably to permit the correlations which Federighi and Fraser attempted, though the variations which Federighi encountered are not unexpected. There is reason to believe that with Urosalpinx, as with other animals, mutations and natural selection are playing a role in the creation and perpetuation of characters which, even though not conspicuous, will vary from one environment to another.

Figures available on the sizes attained by the oyster drill in different geographic regions are summarized in Table 7. These data further serve to emphasize the range in variation in the sizes reported for this species, and thus the difficulty at the present time of assigning average or maximal dimensions to the species as a whole. Table 7 tentatively indicates that three different drill sizes, based on the maximum heights reported, exist: the giant American form of maximum height 51-61 mm., the English form, 39-43 mm., and the small widespread American form, 26-40 mm. whose height range overlaps that of the English form. The average maximum height for the small American form is approximately 33 mm. It should be stressed that to be most useful and meaningful records of drill size should be based on both age and sex groups for each geographic region.

Growth rate and quality of food

Engle (1942) under controlled laboratory conditions in Long Island Sound demonstrated that considerable variation in the rate of growth occurred among four different groups of drills each of which was fed exclusively on an excess of one of the following living animals: oyster (Crassostrea virginica), soft clam (Mya arenaria), edible mussel (Mytilus edulis), and barnacle (Balanus sp.). Observa tions were extended for 13 months, although feeding occurred only between May and November. In the course of the feeding season Urosalpinx grew fastest on a diet of soft clam, less on oyster, even less on barnacle, and least on mussel. Older stages of the soft clam are not ordinarily available to the drill in nature because they are buried in the bottom. When exposed, soft clams are more vulnerable to attack by drills than the other food organisms because of the exposed soft parts, and the fact that drills grew fastest on this diet, although it may reflect a more nutritious diet, may also suggest a more accessible food. Engle further noted that the maximum rate of growth on each of these foods occurred at different periods during the deason: drills feeding on mussels grew most from June 12 to July 12; on soft clams, from July 12 to August 10; and on oysters and barnacles, from August 10 to September 6. This seasonal variation in growth may be associated with a parallel variation in the nutritive value of these food organisms. Engle observed that on the mussel bed originally inhabited by the experimental drills only a few drilled mussels were encountered, while oysters and barnacles there were attacked in large numbers.

TABLE 7. Maximum Height of Urosalpinx cinerea as Reported by a Number
of Investigators for Various Geographical Regions

Maximum Height in mm.	Sex	Total Number of Drills Measured	Region	Source
43	Female	1,700	River Blackwater, England	Cole, 1942
39	Male			
37	Mixed	30,903	Woods Hole, Mass.	Walters, 1910
34	Mixed	?	Woods Hole, Mass.	Pope, 1910-11
29	Female	1,121	Woods Hole, Mass.	Federighi, 1931b
26	Male			
31.1	Mixed	854	Milford Harbor, L. I. Sd., Conn.	Engle, pers. com.
35	Mixed		Little Egg Harbor, N. J.	T. C. Nelson, 1922
37	Mixed	600	Maurice River Cove, Delaware Bay, N.J.	J. R. Nelson, 1931
40	Mixed	300,000	Delaware Bay, N.J.	Stauber, 1943
61	Mixed	-	Seaside, Va. (Giant form)	Galtsoff et al. 1937
51.5	Mixed		Chincoteague Island, Va. (Giant form)	Henderson & Bartsch, 1915
51.2	Mixed		Accomac, Va. (Giant form)	Baker, 1951
33	Female	several hundred	Hampton Roads, Va.	Federighi, 1931c
29	Male			
27	Mixed	several hundred	Hampton Roads, Va.	Federighi, 1930a
33	Mixed	several hundred	Beaufort, N. C.	Federighi, 1930a

46

Sexual Maturity

According to available but not always complete information the oyster drill attains sexual maturity at ages varying from one to three years and heights of 13 to 24 mm. in different regions. Pope (1910-11) in Woods Hole, from field observations states that sexual maturity apparently occurs at the age of one year. He found females about 13 mm. in height, the size attained at the end of one season, depositing egg capsules early in June.

On the basis of field experiments in Delaware Bay Stauber (1943) concludes that sexual maturity is not reached at least until 15, and possibly 25, months of age, at a size of more than 15 mm. in height. In April when the temperature of the water had not yet reached 14°C he confined two sets of drills in cages with oysters at the low water line. The height range of the drills in the first cage was 11-15 mm., and in the second cage, 21-30 mm. By August the drills had grown to heights of 16-24 mm. and 23-30 mm. respectively. Although the larger drills oviposited 895 egg cases during the summer, the smaller drills deposited no egg cases. By October 10 one cluster of egg cases was found in the cage of the smaller drills indicating that the larger drills therein had now reached sexual maturity. Because of their small size Stauber states that the smaller drills when confined in April, 1939, were either 9 months old (derived from the early 1938 summer hatching) or possibly 19 months old (from the late 1937 summer hatching). The smallest isolated drill to oviposit in Stauber's other experimental cages measured 16.5 mm. in height

By confining Urosalpinx in cages with barnacles in English waters Cole (1942) determined that females start depositing eggs when not much smaller than 24 mm. in height at an approximate age of two years, but the number of egg cases oviposited is slight. At the start of the breeding season he noticed that the majority of females 22 mm. or less in height contained undeveloped gonads and immature accessory reproductive organs. From these data and field observations he concludes that full spawning in English waters does not begin until drills are three years old. The differences reported in the age and size of drills at sexual maturity in different regions may express not only incomplete information but the existence of different geographic races.

Longevity and Mortality

Cole (1942) from his size frequency curve and growth mark studies concluded that the approximate duration of life of U. cinerea in English waters is 10 years, and that occasionally drills as old as 13 or 14 years are encountered.

Among the numerous very large females above 41 mm. in height which he examined only two or three contained exhausted gonads, while those up to 41 mm. were observed depositing egg cases. Since the latter, according to Cole, were probably not less than 10-12 years old, the reproductive life of the female, ignoring the first two years of life when a few cases may be deposited, may extend over 7 years. Pope's (1910-11) earlier estimate of three years for the length of life of the drill in Massachusetts may be too low.

A few observations have been reported on the rate of mortality among oyster drills. Stauber (1943) encountered percentages as high as 90% of empty drill shells on vacant bottom in Delaware Bay. Most of these empty shells were inhabited by hermit crabs and undoubtedly represented accumulations over a period of years.

Stauber also reported an accelerating death rate of drills during the winter on grounds on which drill dredging had been performed periodically throughout the winter and spring at water temperatures below which drill migrations occur. On one oyster bottom the mortality rate increased from 5% in December to over 30% in March. Stauber (pers. com.) concluded that the increase in mortality resulted at least in part from the action of the dredge in dislodging and exposing to predation those hibernating drills which remained on the bottom. This is corroborated by observations in England which disclosed that Urosalpinx can withstand unusually cold winters (Orton. 1932) and others in America that at low water temperatures (especially below 5°C) drills become progressively more sluggish and slow in righting themselves after dislodgment (Carriker, 1954).

Food and Food Procurement
Food preferences

Although the oyster drill displays some discrimination in its choice of food, it feeds upon a wide variety of animal species: its own kind, slipper limpets, edible and ribbed mussels, soft and hard clams, scallops, oysters, small crabs, the carrion of fish, and on such lower invertebrates as encrusting bryozoans (Pope, 1910-11; Federighi, 1931c; Haskin, 1935; Galtsoff et al., 1937; Engle, 1940; Carriker, 1943, 1951). On the whole its diet appears to consist principally of small oysters, edible mussels, and barnacles when these are available (Galtsoff et al., 1937; Cole, 1942; Stauber, 1943). The effect of the relative abundance and accessibility of food species on the selection of prey is poorly understood, but it may be conjectured that these factors also influence the diet of the drill.

Among the burrowing bivalves the younger dissoconch stages are attacked when they are partially exposed, which occurs commonly during the byssal stages. Urosalpinx does not generally burrow during the warmer months of the year when

48

feeding occurs. Kellog (1901) reports some destruction in the field of young Mya arenaria, and Gibbs (pers. com.) in Palmer River, Rhode Island, and the writer in Little Egg Harbor, New Jersey encountered significant quantities of drilled young hard clams (Mercenaria (=Venus) mercenaria).

The pronounced predilection of Urosalpinx for the younger stages of the oyster has been emphasized repeatedly (Pope, 1910-11; J. R. Nelson, 1931; Federighi, 1931c; Galtsoff et al., 1937; Cole, 1942; Haskin, 1950), and where oyster culture is practiced intensively the drill feeds almost exclusively on these (Stauber, 1943). Pope and later workers write that when small oysters are placed in the vicinity of large oysters being attacked by drills, the drills invariably desert the larger for the smaller prey.

A number of observations suggest that under some, although not all, conditions. Urosalpinx feeds on the mussel Mytilus edulis in preference to the oyster Crassostrea virginica. Haskin (Galtsoff et al., 1937... from Haskin, 1935) during the course of experiments with individual drills in Cedar Creek, New Jersey placed small Mytilus in two field cages with small oysters and drills. In one cage all four Mytilus were destroyed before any oysters were attacked, and in the other cage only one oyster was drilled while three Mytilus were consumed. Further experiments in large scale trapping of drills in Delaware Bay in which separate traps were baited with young oysters, young mussels, and empty shells, demonstrated that mussels were twice as efficient in attracting drills as oysters. Later (1950) in field cages and in the laboratory when Haskin confined mussels (size not given), young oysters, and drills, he observed that in the field 3.6 and in the laboratory 3, oysters were drilled for every mussel drilled. Cole (1942) confined English drills with a number of food organisms in field cages and observed that drills offered Ostrea edulis spat and Mytilus edulis destroyed practically all the spat before attacking the mussels, which were then quickly consumed. Barnacles were more attractive than mussels, and about equally attractive as one year old oysters.

In feeding experiments in an aquarium Galtsoff et al. (1937) observed that barnacles were readily attacked by drills and that penetration of the prey was effected through the soft parts between the plates of the barnacle. They conclude that drills exhibit a decided preference for barnacles probably because of the vulnerability of the prey.

Under laboratory conditions Urosalpinx feed not only on animals which they have drilled, but also on flesh removed from these animals (Federighi, 1931c). Federighi placed the excised tissues of freshly killed oysters (Crassostrea), clams (Mercenaria) scallops (Aequipecten), oyster drills (Urosalpinx), slipper limpets (Crepidula), pin fish (Lagodon), spots (Leiostomus), and croakers (Micropogon)

49

at one end of an aquarium containing aerated sea water. At varying distances from these meats he positioned drills and watched their movements. The results indicate that under the conditions of the experiment oyster meat was preferred to any other and mollusks were preferred to fin fishes. By introducing living oyster spat he determined that a large proportion of the drills were attracted to the spat in spite of the fact that they had to circumvent freshly killed oyster meat to reach them

No information is available on the role of firm surface encrustations (sessile protozoans, coelenterates, bryozoans, minute algae, and similar organ isms) in the diet of the oyster drill. The writer has observed drills rasping the surface of encrusted shells in the laboratory a number of times. It is possible that this source of food, particularly in the absence of other kinds, may be utilized more than has been appreciated, and may tide them over long periods of time. Since so far as is known drills remaining on grounds from which oysters have been harvested are not eliminated by the removal of the bivalves, it would seem that another source of food is utilized. Stauber (1943) suggests that even a small quantity of food (kind not specified) remaining on the bottom will maintain them. A study of this aspect of the nutrition of Urosalpinx would undoubtedly provide useful information.

T. C. Nelson (1923) in Little Egg Harbor. New Jersey, made the significant observation that where Brachidontes recurvus were attached to young oysters, Urosalpinx continued to feed on the oysters until the oysters developed shells thicker than those of the mussels, then moved to the mussels. The writer (1951) in a series of field cage experiments in this same bay likewise noticed that the presence of such buffer species as thinner valved mollusks affords some temporary protection to the thicker shelled bivalves. In a mixture of 200 Volsella (=Modiolus) demissa, 2-8 cm. long, larger sizes predominating; 74 oysters, 2.5-15 cm. long; 50 Mercenaria mercenaria, 1.5-3 cm. long; and 100 Urosalpinx, 20-30 mm. in height, confined for 47 days during the summer, Urosalpinx drilled all of the ribbed mussels, half of the large and all of the spat oysters, and none of the hard clams. In view of the fact that during the summer the siphonal tip of the valves of Mercenaria frequently projects a short distance above the bottom, it is surprising that no Mercenaria were drilled. It is possible that Mercenaria dislodges such predators by digging deeper into the sediment

It is a provocative fact, and one worthy of further research leading to possible control of the drill; that Urosalpinx only infrequently attack the jingle shell Anomia, even though in many areas within its range this bivalve is one of the most conspicuous and abundant animals on the bottom. In aquaria Galtsoff et al. (1937) observed that of all the foods offered to drills the jingle shell remained untouched. Even drills starved for almost three months spurned them. When these same drills were offered a shucked Anomia, they consumed it within 17 hours. Engle (1942) reports a similar occurrence. These investigators conclude that Anomia shell affords a barrier to Urosalpinx. Loosanoff (pers. com.) finds that Anomia can be drilled by Urosalpinx but only under abnormal conditions when starved drills display interest even in forms which they normally do not attack. An occasional drilled Anomia valve has been found in the field (J. R. Nelson, pers. com.; Andrews, pers. com.). In the early fall of 1953 the writer placed 100 adult drills on a square foot of bottom in a salt pond in Gardiners Island, New York, among oysters of various ages and 12 Anomia simplex about one inch in diameter. These were confined under a screen cage. In late November it was found that one Anomia was partially drilled, and two others were completely perforated, but all three were living. Microscopic examination of the latter showed that the drills had rasped away some of the flesh but that Anomia had regenerated the mantle and secreted a thin layer of shell material over the perforation. Only one instance of unusual predation of Anomia by the drill has ever been reported (Glancy, 1954). This took place in a dense population of large sea stars (Asterias sp.), oyster drills, and jingle shells in Peconic Bay, Long Island. No explanation for this unusual behavior is available. These incomplete observations suggest that the shell material and possibly the flesh of living Anomia may possess a quality which in the majority of cases tends to repel oyster drills.

Drilling behavior

Pope (1910-11), Federighi (1931c), and Stauber (1943) report that in the field the right or flat valve of the oyster is usually perforated since this is generally uppermost, but that in dense clusters or when oysters are resting on their sides, drilling may occur through either valve. Since in native surroundings the left valve of young oysters is usually cemented directly to the substratum and older dislodged single oysters tend to lie on the left valve on the bottom, the right valve is vulnerable to attack. Federighi reports that Urosalpinx generally chooses the uppermost valve of an oyster for drilling even in oysters set in aquaria with both valves exposed.

In a series of detailed plottings of the distribution of the perforations by Urosalpinx in the shells of oysters collected in the field and in the laboratory, Pope (1910-11) discovered that perforations are universally distributed over the entire surface of the shell, and that for the most part the middle areas of the valves are most frequently the site of drilling. He found no evidence to indicate, as had been suggested by earlier observers, that perforations are confined to the limits of the adductor muscle of the oyster, or that the drill always selects depressions or the thinnest portion of the shells of its prey for attack.

51

Federighi (1931c) in a confirmation of Pope's work, found drill perforations in both shells of oysters and on every portion of the valves. Seventy-three percent of the holes were located over or near the place of muscle attachment. In 1919 T. C. Nelson (pers. com.) attached brass tags to 1,000 oysters in Little Egg Harbor, New Jersey, through small holes drilled in the umbones by means of a dentist hand drill applied with constant pressure. He found that the variation in time necessary to drill through the umbones of different oysters varied from a few seconds to five minutes, and interprets this as due principally to the variation in the amounts of prismatic shell and chalky shell present in the umbone. He adds that shell crystals may be harder when attacked on the ends than from the sides. Haskin (1935) using a hand drill and constant pressure found little or no correlation between ease of penetrating an oyster shell and oyster size, and that the oyster drill does not necessarily attack the thinnest shelled oysters nor the weakest spot on the shell of a single oyster.

Observations by Cole (1942) of oyster drills over 25 mm. in height confined with Mytilus edulis in the River Blackwater, England, show that these Urosalpinx drilled principally near the thin edge of the mussel shells. The writer (1951) observed that Urosalpinx appeared to express no specificity of location in drilling Volsella demissa when caged with them in the field.

The actual rate of drilling through shell (as contrasted to the rate of destruction of prey) is probably dependent upon the size of the drill, the relative quantities of prismatic and chalky shell present in the shell of the prey, the temperature and salinity of the water, and other unknown factors (Engle, 1940; Federighi, 1931c; Galtsoff et al., 1937; T. C. Nelson, pers. com.). In 36 experiments in Hampton Roads, Virginia, Federighi demonstrated that the average rate of drilling through oyster shell was approximately 0.4 mm. per day. Galtsoff et al. and Engle record a rate of 0.5 mm. per day. Pope (Field, 1924) determined that the time required by drills to perforate oysters 1.5 inches long is approximately 2 days; 2.5 inches long, 4 days; 3.5 inches long, 6 days; and 4 inches long or longer, 7 days. Orton (1927) estimated that the average time taken for Urosalpinx to drill through an English brood oyster 1-2 inches long was 5-6 days.

The diameter of the hole drilled by Urosalpinx is related to the size of the snail (Stauber, 1943). Because of the bevelled or tapered shape of the perforation it is larger in diameter at the surface of the shell of the prey than at the internal surface. In laboratory observations Stauber found that newly hatched Urosalpinx 1 to 1.5 mm. in height rasped holes ranging from 120 to 220u in outer diameter. Drills averaging 25 mm. in height drilled holes averaging 1.4 mm. outside diameter and 0.78 mm. inside diameter; drills under 15 mm. produced openings 0.8 mm. and 0.54 mm. in diameter respectively.

In a study relating depth of perforation to length of shell of oysters, Pope (1910-11) observed that holes 0.5 mm. deep were confined to oyster shells not exceeding 3.8 cm. in length, and perforations 1.0 mm. and over were limited to valves not exceeding 5.1 cm. Holes 3.0 and 4.0 mm. deep were found in the vicinity of the adductor muscle in oysters ranging in length from 3.5 to 4.4 cm.

Apparently most rapidly growing oysters under 11.3 cm. in length are not immune to the attacks of oyster drills (Pope, 1910-11), although larger so called "dumpy" or thick shelled slowly growing oysters go unharmed. Pope noted that whereas in the laboratory drills perforated oysters as large as 11.3 cm. in length, he found no drilled oysters in the field larger than 7 cm. in length. He states that the greatest mortality in the field occurs in oysters with an average length of 3.2 cm. Since the gaping valves of smaller dead oysters are more readily lost it appears that Pope failed to notice the great rate of destruction that occurs among smaller oysters. Newcombe (1941-42) relates that in the laboratory in no case did small drills kill medium or large oysters. In one instance drills 21 mm. in height did not attack oysters averaging 8.5 cm. in length. Federighi (1931c) and Stauber (1943) say that occasionally oysters over 8 cm. in length are drilled, probably by larger drills. In laboratory investigations Stauber watched newly hatched drills, 1.0 to 1.5 mm. in height, successfully attack oyster spat up to 4 mm. in length. He adds that in the field drills under 15 mm. in height can successfully drill oysters at least 2.5 cm. long. Cole (1942) in laboratory observations found that drills 3.5 mm. in height readily penetrate English oysters ranging in diameter from 3 to 7 mm.

Urosalpinx attack their prey with considerable pertinacity. (Pope 1910-11) states that they frequently return to perforations which they begin. Orton ('1930) describes a case in which a drill resumed its position at a drilling site after four interruptions during which it was removed a few centimeters from the perforation. According to Federighi (1931c), Galtsoff et al. (1937), and Andrews and McHugh (pers. com.), more than one drill may attack an oyster at the same time. Even if one drill succeeds in piercing the oyster shell before another, the unsuccessful drill continues rasping. A maximum of four distinct perforations on one oyster shell is reported. In the field Pope found very few oysters in which the perforation was incomplete; he attributes the cessation of drilling to interruption by oyster cultural operations. Andrews and McHugh (pers. com.) observed a number of instances where a drill missed a living oyster and drilled instead some distance into empty shell beneath.

In the laboratory Urosalpinx continues to drill uninterruptedly throughout the daily (diel) cycle, apparently unaffected by the alternation of day and night (Pope, 1910-11).

Feeding behavior

According to Pope (1910-11) it is not uncommon for two Urosalpinx to extend their proboscides through a drill perforation and thus feed side by side. After the valves of an oyster gape open as a result of drill predation, other drills, attracted by the food, crawl within the opened oyster and literally bury themselves in the meat as they feed. The drill primarily responsible for the killing invariably continues to feed through the perforation. In addition to Urosalpinx, crabs and fish are also attracted to the wounded oyster, and consume it before the drills are able to finish it.

In the laboratory drills have been observed to feed continuously on oyster meat for periods ranging from 12 to 20 hours (Carriker, 1943). Pope (1910-11) observed that an isolated drill in a few days drilled and consumed five young oysters not exceeding 1.9 cm. in length. Galtsoff et al. (1937) write that in their laboratory a single drill penetrated and ate an oyster 5 cm. long; and another devoured about 0.4 cubic inch of oyster meat in 24 hours.

Effect of drilling on oysters

Federighi (1931c) performed experiments in the laboratory which suggest that Urosalpinx while perforating an oyster secretes a toxic substance which kills the bivalve. Oysters penetrated to the surface of the mantle lying against the shell recovered. But those in which the adductor muscle, the pericardial cavity, or the visceral mass were penetrated deeply opened immediately and did not recover. Oysters perforated in less vulnerable areas along the periphery, although they did not gape for several days, also died. In striking contrast oysters similarly perforated with a machinists twist drill continued to live indefinately. No confirmation of these provocative observations has ever been reported. In Federighi's experiments it is of some significance whether the mach inlets twist drill injured the oyster tissues to the same degree as the rasping of the Urosalpinx radula. This was not reported.

Rate of destruction of prey

Numerous records attest the destructiveness of U. cinerea. The available quantitative data on the rate of destruction of oysters has been tabulated in Tables 8 and 9.

Pope (1910-11) stresses the fact that the drill is destructive as soon as it hatches from the egg case, and owing to the diminutive size of both newly hatched drills and recently set oyster spat the real extent of the mortality of very young oysters is generally unknown and unrecognized. Stauber (1943) in Delaware

54

TABLE 8. The Rate of Destruction of Oysters of Various Sizes by Urosalpinx cinerea of various Heights as Reported b Number of Investigators

Average No. Oysters destroyed per drill per week	Length of Oysters	Height of Drills	Number of Drills	Season	Site	Region	Source
33.4	Small spat	2-4 mm.	12	Summer	Lab	Annapolis, Md.	Engle, 1953
33.6	2 mm.	3 mm.	2	May-Nov.	Lab	Milford, Conn.	Engle, 1940
9.7	3-7 mm.	3.5 mm.	20	Summer	Lab	Conway, England	Cole, 1942
2.92	1 yr.	23-36 mm.	97	June-Oct.	Field	Conway, England	Cole, 1942
0.23	2 yrs.	"	42	"	"	"	"
0.84	1 in.	16 mm.	?	May-Dec.	Lab	Annapolis, Md.	Engle, 1953
0.93	"	18 mm.	"	"	"	"	"
1.03	"	20 mm.			"	"	
0.78	"	22 mm.			"	"	
1.09	"	24 mm.			"	"	"
1.14	"	26 mm.			"	"	
1.0	1 yr.	adult	45	May-Nov.	Lab	Milford, Conn.	Engle, 1940
0.56	seed	under 15 mm.	?	Apr.-Nov.	Field	Delaware Bay, N. J.	Stauber, 1943
0.16	"	15.9-18 mm.	"	"			"
0.24	"	over 20 mm.					
0.05	53 mm.	21 mm.	85	Summer	Lab	York River, Va.	Newcombe, 1941
0.14	48 mm.	26 mm.	5	"	"	"	"
0.35		29 mm.	10	"	"	"	"
0.34	10-60 mm. 1 yr.	adult	20	Summer, 9 wks.	Field	Barnegat Bay, N. J.	Haskin, 1935
0.67	26 mm.	adult	1	Summer	?	Eastern Canada	Adams, 1947
0.33	51 mm.	"	1	"	"		"

55

Bay encountered drilled oyster spat as small as 0.36 mm. in diameter which had not yet produced dissoconch shell. He corroborates Pope's conclusion that large numbers of young spat may be drilled, and this may in part explain the apparent disparity frequently encountered between plankton counts of oyster larvae, counts of newly attached oyster spat, and subsequent densities of seed oysters According to Cole in England, where Urosalpinx has become a serious menace, much of the damage it inflicts on the oyster beds passes unrecognized since the shells of small spat up to thumb nail size, one of its chief prey, are easily swept off the cultch and broken up. He states that there is evidence that the drill annually destroys very large numbers of tiny spat within a few weeks of settle ment. In plunger jars in the laboratory he determined that drills averaging 3.5 mm. in height may destroy 9.7 oysters varying in diameter from 3 to 7 mm. per drill per week at a temperature of 20°C. Engle (1940, 1953) also demonstrated a high rate of destruction of very young spat by young drills.

Table 9. The Relation of Water Temperature and
Drill Size to the Rate of Destruction of
Oysters by Urosalpinx cinerea
(Modified from Engle, 1953)

Average Number of 1 inch Oysters Destroyed per Drill per Month of the Following Heights			Temperature Range °C
16-20 mm.	21-24 mm.	25-27 mm.	
1.3	1.6	2.1	13.2-16.2
2.4	3.7	4.9	16.2-18.5
3.3	4.4	5.2	18.5-22.5
---	7.8	11.1	22.5-23.5

A number of estimates on the seasonal damage imposed by Urosalpinx have been reported. Federighi (1931c) from data collected in Beaufort, North Carolina, reports that one drill can kill from 30 to 200 oysters in a season depending on their size. Galtsoff et al. (1937) state that in New Jersey a single drill can kill over 300 very young spat per season, or "...may devour on the average 0.34 adult oysters per week   " The data in the quotation refer not to adult oysters but to oysters one year old (see Table 8) and are taken (no citation given) from Haskin's (1935) work in Cedar Creek, New Jersey. Cole (1942) estimates that in England during a seasonal feeding period a drill feeding exclusively on one year old oysters would destroy 59 spat. Stauber (1943) estimates that a medium sized drill destroys approximately 20 one year old oysters per season

The rate of destruction of prey by Urosalpinx depends, among other possible factors, upon the size of the drill and of the prey, the temperature and the salinity of the environmental water, and possibly the state in the reproductive cycle of the drill. The reported information on the influence of size of predator and prey on the rate of predation is summarized in Tables 8 and 9. These figures highlight the fact that as Urosalpinx increases in size it is capable of destroying greater quantities of oysters of a given size (Newcombe, 1941-42; Engle, 1953), and that a given size of drill can destroy a greater quantity of small oysters than larger oysters (most of Table 8, more specifically Cole, 1942; Adams, 1947). Stauber's (1943) figures seem to indicate that small drills destroy more seed oysters than do large drills; however, seed oysters vary considerably in size, and it is not known what size oysters were fed to the respective size groups of drills.

That seasonal temperatures strongly influence the rate of destruction has been reported by a number of investigators. Pope as early as 1910 stated that drills consume little food during the cold months of the year. He encountered considerable difficulty in attempting to induce drills to feed in the course of ex perimental observations after September in Massachusetts. Galtsoff et al. (1937) write that during 9 weeks in a study (in Cedar Creek, New Jersey; Haskin, 1935; no citation given) the temperature of the water rose from 19.0 to a peak of 28.0°C and dropped to 24.9°C; during this time the average number of oysters destroyed each week in experimental baskets at 10 different stations by a total of 200 drills increased with the temperature from an average of five at the beginning to 8 during the weeks of high temperature and down to 6 at the end of the 9 weeks. Engle (1940) working with drills in an outdoor tank in Connecticut noted that the active feeding season lasted from late May to late November. The peak of feeding occurred during the latter part of July and in August. Cole (1942) in a study of the potential damage of U. cinerea in England confined drills with an excess of oysters in cages in the field. In a plot of the daily rate of destruction of one year old English oysters and water temperature for the season of 1941, Cole demonstrates strikingly that drills start feeding actively as the temperature of the water rises above 14°C, feed voraciously during the major part of the breeding season through mid July, and then a fairly good general correlation occurs between water tempera- ture and rate of feeding during which the feeding rate declines as water temperatures drop. During the active spawning period he obtained destruction rates as high as 6.30 oysters per drill per week, a rate considerably higher than the average figure of 2.92 for the season (see Table 8). These data prompted Cole to warn against the use of destruction rates obtained over a short period of the feeding season in calculating predation rates for the entire season. Pope (1910-11) also mentions that drills continued to feed during the spawning period in his observations. The feeding season in eastern Canada according to Adams (1947) lasts 18 weeks. It begins in the spring when the temperature of the water reaches 15°C and terminates in the fall as temperatures drop to 12.8°C. Engle (1953) using drills collected in the vicinity of Milford, Connecticut, carried out a number of detailed laboratory studies relating the destruction rate of one inch oysters by drills of various sizes to temperature in salinities ranging from 22 to 27 o/oo (see Table 9). He shows clearly that within each size group of drills the rate of destruction of oysters increases noticeably with an increase in temperature. This is also reflected in figures for the average destruction of oysters per drill per month for all sizes of drills from 16 to 26 mm. in height over the feeding period from May to December, respectively: 0.9, 3.7, 4.9, 9.9, 4.5, 2.0, 0.2. In all size groups within this range the greatest destruction rate took place during the month of August.

Andrews and McHugh (pers. com.) at Gloucester Point, Virginia, contribute further information on the influence of seasonal temperatures on the rate of destruction of young oysters. One hundred and twenty four drills ranging in height from 5 to 17 mm. were placed with caged oyster spat 10 to 15 mm. in diameter on January 20, 1954. During the following months the average number of oysters destroyed per drill per week was recorded as follows: January 20-March 10, at an average temperature of 6.5°C, 0.06 oyster; March 10-April 15, at 11.5°C, 0.23 oyster; and April 15-May 20, at 18.5°C, 0.19 oyster. That less oysters were consumed during the last period than during the middle period is explained by the fact that the oysters continued to grow larger during the observations and that supplementary food in the form of barnacles and other organisms set in the cage.

Mackin (1946) has shown that in Virginia the rate of predation of oysters by Urosalpinx also appears to be influenced by the duration of exposure in the intertidal zone. In Finney Creek where oyster spat set abundantly as high as three feet above low water mark on vertical frames, he found that 83% of the spat between -2 and -1 feet below mean low water were drilled; between -1 and 0 feet, 75%; between 0 and 1 foot above low water, 35%; and above this level no drilling occurred. In other areas slight drilling was detected up to two feet above low water mark. Chestnut and Fahy (1953) in a comparative study of the vertical distribution of oysters and of oyster setting in five different localities in North Carolina observed that setting below low water occurred far in excess of that above low water and in general increased in intensity bottomward, but that drilling of these spat by Urosalpinx was not intense nearest the bottom and diminished in intensity off the bottom. Chestnut and Fahy conclude that the high rate of mortality of young oysters below low water level may offer a partial explanation for the peculiar distribution of adult oysters principally in the intertidal zone in this area.

The literature on the culture of the oyster is replete with more or less quantitative accounts of the destruction of oysters by U. cinerea. A chronological review of a number of these serves to emphasize the historical significance and the current magnitude of predation by this snail.

In the last century, as early as 1874 Verrill and Smith wrote that in brackish waters in Vineyard Sound, Massachusetts, the drill was the worst enemy of the oyster, sometimes so numerous as to do very serious damage. Rathbun (1888) noted that Fields Point and Bullocks Cove, Providence River, Rhode Island, were overrun with drills which destroyed fully 95% of the oysters on beds that formerly were the most productive in the area. Rowe (1894) estimated that in southern New England the damage caused to oysters by drills was approximately one million dollars annually. In 1895, 6,000 bushels of two year old oysters introduced in the Shrewsbury River from the Raritan River, New Jersey, were totally destroyed by drills (Bur. Stat. N.J., 1902)

During the present century such reports have greatly increased in number and in detail. Pope (1910-11) in samples tonged in Buzzards Bay, Massachusetts, in August, 1910, found that 87% of the oysters in a sample of 390 oysters were drilled; in another sample of 548 oysters, 87% were drilled. These figures do not take into consideration the losses due to destruction of very small oysters whose valves soon become dislodged and escape detection.

T. C. Nelson in 1922 wrote that in Delaware Bay, New Jersey, mortality of young oysters due to drills reached 60% by October; and estimated that in this bay annual damage caused by the drill to the oyster industry was in excess of a million dollars (1923; J. R. Nelson, 1931).

Federighi (1931c) reports that in Hampton Roads, Virginia, when his investigations on the drill were begun, planters estimated the loss of as many as 90% of their oysters to drills. In a survey of this area Federighi found that 10% of the oysters on cultivated oyster grounds, and approximately 2% on natural rock were drilled. Unfortunately he did not record the size of the oysters drilled.

In experiments with drill traps in Delaware Bay, Galtsoff et al. (1937) noted that as many as 80% of young spat on one year old oyster bait were quickly destroyed. They state that there are many localities in Long Island Sound, New York, and in Chesapeake Bay, where drills commonly kill 60 to 70% of the seed oysters present, and not infrequently destroy the entire crop (also Engle, 1940); and estimate that on the eastern shore of Virginia between Chincoteague and Cape Charles the probable annual loss of oysters to drills was about $150,000.

Stauber (1943) relates that a ground in Delaware Bay was planted heavily with small oyster seed in May, and during the summer additional oyster spat catches amounted to 49 spat per shell. Although drill trapping was begun soon after spatting began, most of the summer spatfall was destroyed and the original spring planting was much reduced by drills. In a second instance in the same bay an oyster farmer obtained a heavy spatfall of oysters (91.6 spat per shell) on cultch planted on an isolated ground. By the end of October at least 65% of the spat were drilled, and soon after most of the set was a total loss.

Newcombe and Menzel (1945) report that in 1944 more than 40% of the oysters on Nansemond Ridge, mouth of the James River, Virginia, were killed by drills; and that on the Sea Side it is not uncommon to find a 70% mortality

Mackin (1946) observed that 75 to 83% of the oyster spat in Wachapreague, Virginia, were destroyed by Urosalpinx between June and December.

In eastern Canada on one heavily drill infested oyster ground at Malagash, 32% of the oysters present were destroyed in one summer

Cole (1951) notes that in Essex rivers, England, approximately 75% of English oyster spat present are destroyed during the first year of life.

Mistakidis (1951) found an average of 1.75 drills per square meter on an English oyster bottom in a poor state of cultivation, and using Cole's (1942) figure of 0.9 spat destroyed per drill per day, has calculated that during a period of three months in the summer this concentration of drills feeding at this rate on a 15 acre ground would destroy a maximum of 8,840,000 oyster spat.

Engle (1953) reports a case in Tangier Sound, Maryland, at a time when drills were abundant and salinities high in which 50% of the seed oysters planted in April were destroyed by drills by July, and 100% by October of the same season.

At the Institute of Fisheries Research pier, Morehead City, North Carolina, Chestnut and Fahy (1953) found that 23 spat in a total of 516 spat collected on experimental cultch near the bottom were drilled by Urosalpinx.

RELATION TO ENVIRONMENTAL FACTORS
Substratum

Studies in which the distribution of the oyster drill has been related to the nature of the substratum appear to agree rather consistently that soft muddy bottoms devoid of shell, stone, living epifauna, and other hard objects are unfavorable for

60

its growth and multiplication, and that it does not occur in any significant quantity on these bottoms (Federighi, 1931c; Stauber, 1943; Adams, 1947; Cole, 1951). Although Federighi states that the drill is unable to cross muddy areas, Adams cites an example in Malagash Basin, Canada, where they traversed a barren mud bottom about 75 feet wide and reached a dense concentration of oyster spat which were being reared in a dyked area. Engle (1935-36) in laboratory aquaria at temperatures 8°C or below observed that drills moved at a rate of about 0.087 cm./min. over sandy shelly gravel bottom, but only 0.063 cm./min. over mud bottom overlaid with a few shells. Stauber adds the observation that every several years in Delaware Bay, especially on bottom muddied over for a time, which probably reduces the drill population, oyster spatfalls occurred which grew to market size.

It is also suggested that an unstable sandy bottom devoid of firm objects is unfavorable for locomotion and possibly for survival (Federighi, 1931c; Cole, 1942; Stauber, 1943.). Stauber, corroborating Cole, noticed that where sand predominates and shell or oysters are few, low densities of drills were encountered in field trapping and dredging. The writer has observed in aquarium studies that although drills remain on hard surfaces (clusters of oysters and sides of aquaria) much of the time, they occasionally and quite readily creep over sand or mud from one hard surface to another.

Favorable substrata for U. cinerea seem to consist of lower intertidal and subtidal surfaces of wood, metal, and rock, and of firm sand and mud or mixtures of these overlaid by shell and/or living oysters, mussels, and other epifauna; and combinations of these surfaces (Federighi, 1931c; Stauber, 1943; Adams, 1947; Mistakidis, 1951; Cole, 1951; and others).

Salinity
Minimum survival salinity

U. cinerea frequently inhabit estuaries where seasonal fluctuations in salinity resulting from rain fed floods of fresh water impose rigorous, even lethal, conditions. This is particularly evident in the headwaters of estuaries where an unseen oscillating barrier, the minimum survival salinity for the drill, inflicts a rigid chemical restraint upon upbay distribution (T. C. Nelson, 1922; J. R. Nelson, 1931; Galtsoff et al., 1937; Stauber, 1943; Engle, 1953; Glancy, 1953). Careful studies have shed some light on the response of the oyster drill to variations in salinity in relation to such ecological factors as temperature and time

Federighi (1931c) choosing as the "salinity death point" that locus in his data where approximately 50% of the drills in laboratory trials died after 10 days of exposure to varying salinities at summer temperatures, concludes that drills

61

collected in Hampton Roads, Virginia, in two regions with average summer salinities of 15 and 20 o/oo possess a salinity death point of about 11.7-12.5 o/oo. He explains that the discrepancy between the death point salinity observed in the laboratory and the salinity below which drills were not found in Hampton Roads during the summer, i.e., 15 o/oo, arises from the fact that in the spring salinities in these waters fall as low as 12 o/oo, figures close to the experimentally determined salinity death point of 12.5 o/oo.

In his laboratory investigations Federighi placed 20 to 50 drills in a container for each salinity tested and held the drills under water by means of a wire screen stretched below the surface. The water was not changed during the 10 day testing period. The use of metal screens (kind not mentioned) for this purpose, particularly if galvanized, is highly questionable since such ions are toxic to many aquatic organisms when present in concentrations above low levels, and may explain in part the death rate of drills in some of Federighi's higher salinities. Federighi draws attention to the fact that his experiments illustrate the narrow margin of salinity safety for drills living in relatively low salinities as contrasted to a broader range of non lethal salinities for those in relatively high salinities.

In similar laboratory studies in Delaware Bay, Sizer (1936) showed that at summer temperatures drills survived two weeks in salinities between 15 and 30 o/oo; in salinities of 5, 35, and 40 o/oo all drills died in four days, and in 10 o/oo died a few days later. Engle (1935-36) concluded that salinities above 35 and below 10 do not support drills, and that between 30-35 and 10-15 o/oo the vitality of the drill is reduced. Laboratory experiments by Galtsoff et al. (1937) indicate that a salinity of 11 o/oo kills drills and that this salinity constitutes a natural barrier to the distribution of the gastropod. Water fresher than this killed drills more quickly: in 10 o/oo they died in 7 days; in 5 o/oo, in 4 days.

In an evaluation of earlier work in this field, Stauber (1943) stresses the important fact that the factors of time and temperature were not adequately emphasized, and suggests that instead of Federighi's term "salinity death point" the more accurate term "salinity death time" be substituted. He adds that although Federighi computed his value as that point at which approximately 50% of the drills died in the specified time interval, his data show that a few drills survived the experimental procedure.

The fact that heavy spring rainfall repeatedly reduces salinities in Delaware Bay to low concentrations and for long periods of time, which by all previous information on salinity death times should have destroyed drills over large areas of the bay --- but do not, prompted Stauber (1941; 1943) to undertake a laboratory reinvestigation of this problem. He confined groups of 10 to 15 drills in dishes containing sea water of the desired salinity which he changed daily at first and less

62

frequently later. Stauber conducted his experiments at winter temperatures consistently below 20°C which fluctuated roughly with those out of doors, while previous experimenters performed their experiments only during the summer months at temperatures usually well above 20°C. Stauber soon determined that at winter temperatures drills could survive for considerable periods of time under less saline conditions than previously reported. Drills were able to attach to the sides of the containers in salinities as low as 7 o/oo though they later died under these conditions. Half of the drills were able to attach after 26 days of exposure to 8 o/oo, after 121 days to 9 o/oo, and 248 days to 10 o/oo. Without food one drill was capable of attachment in 8 o/oo on the 136th day, in 11 o/oo on the 234th day, and in 9 o/oo on the 344th day. Two of the original drills in one of these dishes were still alive and capable of attachment and movement after more than 19 months in water of 10 o/oo.

Stauber then instituted a series of 6 experiments to test more carefully the effect of dropping temperatures on drill survival in low salinities. These extended from August to March and were conducted at temperatures roughly paralleling outdoor winter temperatures. Lots of 10 drills each, a total of 60 drills per experiment, were confined in dishes with saline water of 6, 7, 8, 9, 10, and 11 o/oo. The results of these exposures are indicated in Table 10. Though not reported, it is assumed that the temperature of the water in the experiments during the second 30 day period was considerably lower especially in the experiments performed in the late summer and fall. Stauber concludes that at lower water temperatures he obtained not only greater early recovering and attachment of the drills but also greater ultimate survival and resistance to the effects of the unfavorable conditions.

Further experimentation was necessary to demonstrate whether drills would survive conditions duplicating those in nature during low temperatures when periods of low salinities caused by excessive rainfall are followed by periods of higher salinity. At temperatures prevailing outdoors in early spring Stauber placed 50 drills each in large aquaria containing aerated water of the following salinities: 0, 1, 2, 3, 4, 5, 6, 7, 8, and over 15 o/oo. On successive days he removed aliquots of drills and placed them in water saltier than 15 o/oo. The number of drills recovering sufficiently from this double treatment to attach to glass in the aquaria and the time of exposure required to kill the drills in each salinity, are given in Table 11. Stauber's experiments very clearly demonstrate that in the field as well as in the laboratory the rate of destruction of drills by low salinities will be determined by the salinity of the water, the prevailing temperatures, and the duration of the conditions. Stauber (1943a) has introduced a graphic means of presenting salinity in an estuary, based on Delaware Bay, which indicates the variations and extremes which can occur. A recording of these is important in

63

TABLE 10. The Effect of Successively Lower
Water Temperatures on the Survival of
Urosalpinx cinerea in low Salin-
inities (After Stauber, 1943)

Mean Water Temperature during lst. 30 Days of Experiment °C	Minimum Salinity in which Drills Survived 60 Days o/oo	Total Number of Drills (out of 60) Surviving in:	
		14 Days	60 Days
19.3	11	5	5
17.7	10	8	5
17.3	9	20	18
16.7	8	32	24
14.0	9	41	27
10.7	8	47	34

TABLE 11. Rate of Mortality of Drills Subjected First to Salinities Ranging from
0 to 8 o/oo for Successive Days and
then Immersed in Salinities over
15 o/oo at Low Outdoor Spring
Temperatures in New Jersey
(After Stauber, 1943)

Initial Salinity o/oo	Time During which attachment in Initial Salinity was re-Reduced to 50% of the Drills Days	Time in Initial Salinity in which all Drills Died Days
0	9	10
1	9	11
2	9	13
4	9	13
6	9	14
7	10	over 15
8	over 15	over 15
over 15	over 15	over 15

estuaries since "Brackish water species are kept within certain spatial limits by the effects of the extremes of salinity and of their duration, not by the means" (Stauber, 1943a).

Engle (1953) in a further analysis of Federighi's (1931c) data, points out that Federighi records only one case in which a total mortality of drills occurred at low salinities, and in all the other tests 10% of the drills survived the so called death point salinity. In defense of Federighi's method Stauber (pers. com.) notes that the use of the 50% mortality point is the approach of the laboratory physiologist, rather than that of the ecologist to whom 10% survival may mean repopulation of an area. In an extension of the work of Federighi and in an effort to determine the duration of survival of 100% of the drills at various temperatures and salinities Engle (1953) collected drills in waters characterized by a salinity range of 22 to 27 o/oo, and exposed each of groups of drills to a different salinity in a series ranging from 3 to 27 o/oo for a period of about a month in the laboratory in Annapolis, Maryland. In order to determine the influence of temperature on drill survival at each of these salinities, he established the salinity series in quadruplicate and exposed each series to one of the following mean temperatures: 17.5, 18.2, 20.1, and 20.5°C. Some of these results are tabulated in Table 12. Engle discovered that all salinities below 14 o/oo were lethal to all drills within the temperature range of 15.4 to 23.0°C, but that death came more slowly at lower temperatures. All drills in salinities of 16 o/oo or above survived the 30 day experimental period without ill effects. The death time in salinities of 14 o/oo and below increased with increasing salinities and with decreasing temperatures (see Table 12). Experiments at temperatures intermediate between those given in Table 12 produced death times also intermediate between the values given there.

Table 12. Death Rates of Urosalpinx cinerea as
Affected by Temperature and Salinity
(After Engle, 1953)

Mean Temperature °C	Time in Days During Which 100% Mortality of Drills Occurred in the Following Salinities:		
	7.5 o/oo	10.0 o/oo	12.0 o/oo
17.5	13	22	30
20.5	7	9	12

Engle's data also clearly express, within the temperature range of his experiments, the wide range of variation of death times for different drills even at a given salinity. For example, in a salinity of 5 o/oo the first drill died in approximately 4 days, 50% of the drills died in 9 days, and 100% in 11 days; in 12 o/oo the first drill died in 5 days, 50% in 16 days, 100% in 30 days; in 14 o/oo the first drill died in 5 days, 50% in 15 days, and 100% in 49 days. At a salinity of 14 o/oo the relation between temperature and salinity death times, so nicely established at lower temperatures, disappeared: at a mean temperature of 17.5°C, 100% mortality occurred in 18 days, at 18.2 and 20.1 in 25 days, and at 20.5°C in 49 days. Engle concludes that within the temperature range of 15.4 to 23.0°C the minimum salinity tolerance level of these drills lies somewhere between 14 and 16 o/oo, a range intermediate between those given by Federighi for Hampton Roads and Beaufort drills.

Although the designs of the various experiments on the salinity death times of the oyster drill differ considerably, it is demonstrated in some of the experiments and may be inferred from others that at temperatures close to those of summer conditions drills do not generally survive below the following approximate salinities:

> 17 o/oo, Beaufort, North Carolina (Federighi, 1931c);
> 12 o/oo, Hampton Roads, Virginia (Federighi, 1931c);
> 16 o/oo, Long Island Sound, Connecticut (Engle, 1953);
> 12-15 o/oo, Lower Delaware Bay, New Jersey (Sizer, 1936;
>   Galtsoff et al., 1937; Stauber, 1943).

The researches on salinity show conclusively that the variables of temperature and time exert a significant influence on rate of survival: at optimum summer temperatures drill mortality rates increase rapidly as salinities fall, but this rate is markedly reduced as temperatures drop, so that at low winter temperatures Urosalpinx can withstand unusually low salinities for protracted periods. It is further indicated that early conditioning of the drill also influences survival (Federighi, 1931c; Stauber, pers. com.).

Movement

The rate of movement exhibited by Urosalpinx in the narrow salinity zones at either extreme of its normal salinity range has not been satisfactorily explored. Sizer (1936) in laboratory experiments at 24°C made the unrealistic observation that drills were inactive at salinities below 20 o/oo and above 30 o/oo, and that they crept most rapidly at about 25 o/oo or above. Stauber (1943) believes that Sizer did not allow sufficient time for his drills to adjust to the experimental salinities. He questions Sizer's figures on the basis that he collected drills in drill traps in some

67

portions of Delaware Bay during entire seasons when salinities below Sizer's
limits prevailed. Galtsoff et al. (1937) at summer temperatures found drills inactive in Delaware Bay in salinities below 15 and above 29 o/oo, and most active
at salinities near the upper limits of this range. Stauber discovered in his salinity
experiments that drills are able to attach and to crawl in almost every salinity in
which they survive. He concludes that in Delaware Bay at low winter temperatures
they can move in salinities as low as 8 o/oo

Drilling and feeding

Very little is known of the effect of salinity on the drilling rate in
Urosalpinx. Haskin (1935, quoted by Galtsoff et al., 1937, without citation) in
summer field studies in Cedar Creek, New Jersey, found that drills ceased drilling approximately below a salinity of 10-12 o/oo. Above this level the rate of
drilling was not appreciably altered by an increase of salinity up to 25 o/oo.
Sizer (1936) in Delaware Bay noted no apparent correlation between the number of
drills captured on baited drill traps and the salinity, but all this trapping (according to Stauber, pers. com.) took place within the brackets of Haskin's figures for
no correlation.

Growth

Federighi (1930a) in measurements of several hundred drills found that
the height of Urosalpinx in Hampton Roads, Virginia, averaged 21-25 mm., and
that drills in Beaufort, North Carolina, collected in salinities some 10 o/oo higher,
averaged 13-17 mm. These data suggested to him that drills grow to a larger size
in brackish than in saline water. He adds that this difference in size is probably
not due to a difference in salinity alone, since other unknown factors may be the
controlling ones.

In a later paper (1931b) he reports that the average height of drills
gathered in the vicinity of Woods Hole, Massachusetts, in salinities only
slightly lower than those in Beaufort, was 21 mm.; that of drills in England in
salinities approximately 34 o/oo, 29-30 mm.; and that of drills in Maurice River
Cove, Delaware Bay, in salinities of 21-30 o/oo, 28.8 mm. (J. R. Nelson, 1931).
Andrews and McHugh (pers. com.) write that drills collected in the York River near
Gloucester Point, Virginia, are much smaller than those taken at Hampton Bar,
Virginia, though salinities are not greatly different. The writer is in complete
agreement with Federighi's conclusion that the factors which may influence the
degree of size attained by drills are unknown. Fundamental studies of the effect
of various environmental factors on the growth rate of this gastropod are much
needed.

Oviposition

In an effort to determine the influence of salinity upon the rate of ovi-position in Urosalpinx, Federighi (1931c) distributed 13 cages enclosing drills and oysters over a wide range of salinities in Hampton Roads, Virginia. The cages were lost in 5 to 7 weeks, but in this interval Federighi obtained the following average number of eggs per case: 8.9, 8.4, 9.3, and 8.9 in the following average salinities, respectively: 12, 15, 17, and 20 o/oo. He concludes that the oyster drill reproduces wherever it survives, and that a salinity of 17 o/oo seems to be the optimum salinity for the maximum production of eggs per case in these waters. The differences obtained by Federighi in oviposition, however, are too slight to be statistically significant; and on the basis of Stauber's (1943) informa-tion on the relationship between drill size, egg case size, and number of ova per case, Federighi's correlation is of doubtful significance.

Haskin's (1935) field data fail to confirm Federighi and clearly demon-strate that in Barnegat Bay, New Jersey, drills can survive where they cannot reproduce and that in estuarine areas of fluctuating salinities they may be able to oviposit but the ova so deposited may not survive. Stauber (1943) from data accumulated in extensive drill trapping in Delaware Bay, concurs with Haskin, and adds that there is a strong suggestion that salinities close to 15 o/oo are necessary for drill oviposition.

Egg case stages

According to Haskin (1935) post hatched stages of Urosalpinx survive in low salinities in which their eggs fail to develop and in which no oviposition occurs. He noticed that uncleaved eggs placed in a salinity of 7.9 o/oo failed to undergo cleavage and started to decompose in 19 hours. In general eggs cleaved normally in salinities above 14 o/oo even when removed from the egg capsule. Development of egg case stages varying in age from one to five days was also abruptly termin-ated when the capsules were subjected to a salinity of 7.9 o/oo, and within 18 hours all were dead. Haskin concludes that a short period of low salinity, even for the duration of one tide, will probably kill egg case stages up to one week of age. The effect of low salinities on the later protoconch stages was not determined. Sizer (1936) observed that protoconchs continued their development within the egg case over the salinity range of 10-30 o/oo, and when removed from their cases survived in these solutions for a number of days.

Stauber (1943) extended these studies to include all prehatching stages of the drill. Egg cases containing various stages of development were segregated in four containers in which salinities of 5, 10, 15, and over 15 o/oo respectively were maintained for 42 days at prevailing summer temperatures. No stages survived

69

5 o/oo. In 10 o/oo a few shelled veligers were still alive on the 42nd day, but no hatching had occurred. In 15 o/oo normal drills started hatching on the 20th day. In salinities over 15 o/oo normal drills were seen on the 16th day.

These experiments indicate that lower salinities suppress the rate of development particularly of the youngest stages to the point of death, and that salinities higher than 10 o/oo are required for complete development of the egg case stages in Delaware Bay.

pH

Sizer (1936) writes that Urosalpinx is viable over the same pH range in which it moves. In Delaware Bay he obtained no correlation between numbers of drills and pH of the sea water in which they were taken. In the laboratory they remained active over the range of 6.5 to 9.1, surviving at least one week in sea water of these hydrogen ion concentrations. He found some evidence suggesting that the drill moves less rapidly in the more alkaline and in the more acid extremes of this range.

Temperature

The activities of U. cinerea are strikingly influenced by temperature. The "critical temperatures" at which its different biological activities have been stated to begin and cease in a given geographic region may not be as constant for all individuals in a population as has been suggested (Table 13). The thermal limits between which at least some of its functions are carried on seem to vary inherently in different individuals, and possibly in different stages of the life cycle. In addition the range of these limits may be affected differentially by the interplay of associated environmental factors and thus may vary inter- and intraseasonally in the same geographic region. Finally, a number of investigators have demonstrated that in many, but not all, instances the minimum thermal limits above which such physiological functions as locomotory movement, drilling and feeding, and oviposition occur, decrease northward in the latitudinal distribution of the species.

Movement and hibernation

As the temperature of the water drops, creeping rate in the oyster drill gradually decreases until at a low temperature range, which seems to vary among different drills in a given region and in drill populations in different regions, movement ceases altogether. A number of studies by Adams (1947), Cole (1942), Engle (1935-36), Federighi (1931c), Galtsoff et el. (1937), Gibbs (pers. com.),

70

Mistakidis (1951), Orton (1930), Stauber (1943), and the writer (1954), illustrate that as water temperatures descend in the fall oyster drills at least in the northern areas of their geographic distribution exhibit the following general movements: those inhabiting subtidal bottoms crawl off elevated objects onto the surface of the bottom and a number bury in the sediment; some of those in intertidal areas burrow in the bottom and others probably migrate into deeper water. The reverse migration takes place as water temperatures mount in the late spring.

Federighi (1931c) noticed in North Carolina that drills retained in running sea water aquaria maintained at temperatures prevailing outdoors became inactive below 10°C (see Table 13) remaining attached to the substratum or lying passively on the bottom. A temporary rise in temperature above 10°C anytime during the winter stimulated slight locomotory activity and the activity increased as the temperature ascended. Gatlsoff et al. (1937) working in the laboratory in the northeastern states disclosed that in water temperatures ranging from -3.0 to 9.6°C locomotion in drills slowed to a rate of 0.024 cm./min. as the temperature descended to 4.5°C, and completely ceased below 2°C. At temperatures below 2°C drills either hibernated on the surface or buried in the bottom attached to partially buried oysters, shells, and other hard objects, with the tip of the siphon projecting slightly above the surrounding bottom. Loosanoff and Davis (1950-51) in laboratory experiments have found that a few drills in lots from Rhode Island, Long Island Sound, and Virginia extend the foot slightly but do not turn over when on their backs at a temperature of 0°C. At 3°C drills from these regions and also from New Jersey and North and South Carolina are able to turn over. None of these drills was able to attach at 0°C (Table 14). Galtsoff et al. (1937) uncovered no evidence that drills seek and congregate in cavities of empty shells for protection during cold weather. Adams (1947) noticed that hibernating drills attached to oysters and shells became coated with a layer of silt which hides their typical form and makes them very difficult to detect. Engle (1935-36) and Stauber (1943) both observed in the field in New Jersey that not all drills bury in the bottom since some were found deeply wedged in crevices formed by clusters of oysters. Stauber reports that drills migrating off intertidal reefs in the fall were discovered later more or less completely buried in the bottom around the edges of the reef usually clinging to shell and with siphons oriented upward and presumably in contact with the water. Engle (1935-36) writes that in the winter in Delaware Bay the drill dredge collected drills on hard as well as on soft bottom, and that numerous drills were collected with a drill dredge equipped with a scraper bar, indicating that the snails were present on the surface or not far below it

In order to obtain further information on the wintering over behavior of the oyster drill the writer (1954) carried out a series of parallel field and laboratory studies in 1953-54. Oyster drills collected in Shark River, New Jersey, were maintained in an aquarium in a closed running sea water system in which water

71

temperatures approximated those out of doors and on bottom prepared to simulate a natural one in the field. Field observations consisted of a study of the vertical disposition of drills in a natural population in Shark River and of drills caged over bottom in a tidal pond on Gardiners Island, New York.

Field and laboratory observations disclosed that in general during the winter drills remain attached to hard surfaces, and their vertical distribution ranges from those stationed somewhat superficially in the hollows of upturned empty shells to those completely buried in the bottom. Adherence of the drill to the substratum is noticeably weaker at lower temperatures, particularly below 5°C, and considerable variation occurs in the degree of torpor exhibited by different drills.

Laboratory studies revealed that drills burying in the sediment may move backward into deep sediments along vertical hard surfaces or creep forward in shallow sediments but go no deeper than the siphon tip. In every instance observed this remained just above the surface of the bottom in contact with the water. No drill was ever seen in the process of creeping backward or forward into deep sediment, but numerous drills buried to different depths, always with the siphon tip upward, were seen repeatedly, and in no case was the sediment appreciably disturbed immediately around the drills suggesting that the drills had moved into the bottom siphon tip forward and then turned. The question also arises as to whether a drill would block its siphon with sediment if it should move siphon end first into the bottom. It was also observed that in uneven bottom drills frequently bury at the lowest level in depressions and thus may come to lie some distance below the general level of the bottom. At least 75% of the drills buried partly or completely in the bottom during the colder months of the year. Those remaining on the surface generally placed themselves in the hollows of empty shells and under these shells and soon became obscured by silting. Considerable variation occured in the locomotory activity of different drills during the winter. A few started burying in the bottom at temperatures above 10°C. Temperatures below 5°C reduced drill movements onto the sides of the aquarium almost to zero, sometimes after a lag of a few days, and in general suppressed such movement for as long as 10 days even though in the interim water temperatures again rose intermittently as high as 13°C. During the coldest period of the winter when water temperatures fluctuated between 1.6 and 4.0°C, the arrangement of the drills did not change appreciably, except that exposed drills moved a little deeper among the shells and became less evident. Individually marked buried drills also exhibited great, though localized, variation in movement throughout the winter at temperatures approximately above 2°C. Many of those only partially buried moved deeper or horizontally or completely vacated the hibernation site for another, or did not bury again. Drills buried to the siphon tip likewise moved horizontally or upward, and often sought other

72

hibernacula. In general movement within the sediment did not extend over a distance of 1/8 or 1/2 inch. In spite of occasional intervening warm periods buried snails remained stationary for periods varying from a few days to 56 days. In aquarium observations of a preliminary nature of drills from Bogue Sound, North Carolina, the writer observed that at temperatures approximately above 15°C some drills burrowed mole-like over the slate bottom of the aquarium in fine shallow sand. This suggests that these southern drills respond to dropping temperatures by burying at higher temperatures than the New Jersey drills.

The writer's observations corroborate those of Galtsoff et al. (1937) on the initiation and cessation of movement of New Jersey drills at low temperatures, Stauber's (1943) observation that drills for the most part do not creep off the bottom onto elevated objects at temperatures below 5°C, and the observations of these investigators that the drills which bury in the bottom move no deeper than the siphon. Experiments with more carefully controlled stable and fluctuating temperatures and utilizing drills from various latitudinal regions should reveal additional information on the behavior of the oyster drill at low temperatures.

Drilling and feeding

Drilling and feeding in U. cinerea cease as the temperature of the water drops to reported limits which vary not only for a given region but also for different geographic locations (Tables 13 and 14).

In Edge Cove, New Jersey, T. C. Nelson (1922) noted that drills fed during the interval from April 13 to May 16 when water temperatures were rising from 15.6 to 18.3°C. In laboratory observations in Hampton Roads, Virginia, Federighi (1931c) found that drills begin to feed when the temperature of the water climbs above approximately 15°C and ceases when it falls below this point. He also noticed that drilling is completely halted by a sudden drop in temperature. In two cases drills actively feeding in water at 27°C abruptly halted drilling and moved away as the temperature fell suddenly to about 20°C. Federighi's observation led Stauber (1943) to a possible explanation of partly drilled oysters found in cages in Delaware Bay. Galtsoff et al. (1937) state that in Delaware Bay drilling occurs at all temperatures between 9.5 and 28°C. Stauber (1943) confirmed Galtsoff et al., but set the lower limit at 10°C. In addition he hypothesizes that since substances emitted by the oyster stimulate the drill to seek the oyster as food, the lower limit at which drilling begins and ceases in Delaware Bay may not be due to a critical temperature value in the response of Urosalpinx. At temperatures between 5 and 10°C oysters are entering or emerging from hibernation with consequent considerable variation in capacity to attract drills. This might account for the sizeable differences in catches in the traps during this temperature range.

73

In the cages in English waters Cole (1942) noticed that drilling may begin as soon as the temperature of the water exceeds 11-12°C and ceases when the temperature descends below this limit. Engle (1953) found that at Annapolis, Maryland, the first drilling of oysters took place in temperatures ranging from 11.6 to 17.4°C; in Long Island Sound, New York, it began after the water temperature reached 11.6°C; and in Delaware Bay it commenced at 12°C and discontinued at about 8°C. In a review of Federighi's (1931c) data he notes that Federighi's drills did not stop feeding until temperatures fell below 7.6°C. Andrews and McHugh (pers. com.) at Gloucester Point, Virginia, observed that drills 5 to 17 mm. in height drilled oyster spat in the period January 20 to March 10 at an average temperature of 6.5°C during which the highest temperature recorded was 9.5°C. Thus it appears that Federighi's minimum temperature for feeding is probably too high. Loosanoff and Davis (1950-51) using drills from various geographic regions in laboratory experiments found that a few drills from Rhode Island, Long Island Sound, New Jersey, and Virginia attempted to feed at 6°C; at 9°C a number of these drills and a few of those from North and South Carolina fed (Table 14).

Haskin (1935) in field studies in Barnegat Bay, New Jersey, in which 20 drills were caged with 15 one year old oysters in each cage at 13 different stations, noticed that in the range observed, 19 to 28°C, the rate of destruction of oysters increased with the temperature.

Oviposition

The reported temperatures at which the drill begins to oviposit also vary in different geographic regions but the postulated correlation with latitude (Table 13) is not always clear. In both laboratory and field observations in Hampton Roads, Virginia, and in Beaufort, North Carolina, Federighi (1931c) found the first egg cases after the temperature of the water had risen above 20° C for some time. In Delaware Bay, New Jersey, according to Galtsoff et al. (1937) egg case deposition begins after the temperature of the water has reached 13.9°C. Stauber (pers. com.), however, seriously questions the accuracy of this observation; in drill trapping and drill dredging operations in the same bay and confining drills in cages with oysters over a period of 7 years, he (1943) found that in general no egg cases were laid at a temperature lower than 15°C. During July as the water temperature approached 25°C oviposition diminished almost to zero and remained at this low level until September when a period of water temperatures fluctuating from 25 to 15°C initiated a second but minor spawning. He notes that the spawning period of drills in shoal areas will be advanced over that of those in deep water, whereas that of drills in intertidal areas may either precede or lag behind deep water areas depending upon atmospheric conditions. Loosanoff (1953) in Long Island Sound has found clusters of recently deposited Urosalpinx egg cases at depths of 30 feet at a minimum

74

temperature of only 10.9°C. Gibbs (Stauber, 1950) observed cases of oviposition in Narragansett Bay, Rhode Island, in 12 feet of water at temperatures as low as 11.1°C. In England Cole (1942) finds that spawning begins each year when the temperature of the water reaches 12-13°C. Adams (1947) notes that in Canadian waters Urosalpinx spawn at 20°C.

In laboratory experiments with lots of drills from various regions Loosanoff and Davis (1950-51) observed that egg laying began at 12°C in Rhode Island drills, at 15°C in those from Long Island Sound, New Jersey, and North Carolina, and at 19°C in drills from Virginia and South Carolina (Table 14).

Tolerance to extremes of temperature

U. cinerea is apparently an eurythermal organism. According to Sizer (1936) high temperature regions do not prove to be a barrier to its distribution in New Jersey, for it is very active in the shallow waters along the Delaware Bay where temperatures may rise above 30°C for considerable periods of time.

At the other extreme of the temperature range, the drill appears to withstand low temperatures which apparently are lethal to some related drills. Orton and Lewis (1931) and Orton (1932) were able to record the relative abundance of Urosalpinx and of two related drills, Ocenebra and Nucella, and the associated temperatures and salinities in the River Blackwater estuary, British Isles, during the period 1928-1930. During the winters of 1926-1928 and 1930 temperatures of the water did not fall much below 3.8°C. But during most of January and February of 1929 the mean monthly temperature dropped 2.2 to 5.6°C below normal. Two day temperature means for January fell to a minimum of 0.4°C and for February to -1.5°C. In the three year period 10,852 drills were dredged from the lower more saline grounds of the estuary where all three species of drills were normally abundant. Urosalpinx was apparently unaffected by the cold of 1929, and its numbers may have actually increased; Nucella suffered severe losses; and Ocenebra was almost annihilated. Orton lists the sustained low temperatures and associated low salinities which came at the beginning of the cold period as the probable causes of the heavy mortality. He notes that the distribution of Urosalpinx extends farther up the estuary into less saline water than does that of the other two species, indicating that Urosalpinx is also more euryhaline than the related species. Galtsoff et al. (1937) observed Urosalpinx at temperatures as low as -3°C and report no mortalities.

Water Currents

Federighi (1929, 1931c) in carefully executed laboratory experiments making use of a celluloid trough and water currents of the turbulent type, discovered that a

75

current of 1.25 to 7.6 cm./sec. stimulates the oyster drill to turn immediately into and move against it. The speed of the current does not affect the rate of creeping, although in swifter currents more work is performed in crawling. Faster currents stimulate the drill to turn into it more rapidly. Removal of eyes and tentacles does not interfere with the rheotactic response. Stauber (1943) in a simplified flow chamber determined that at current velocities as low as 0.2 cm./sec. Urosalpinx no longer exhibits a positive response to current and may even move down stream. At increased velocities such as those employed by Federighi prompt orientation occurred.

In Delaware Bay both Sizer (1936) and Stauber (1943) discovered in water at least 6 feet deep that drill trapping is no more effective in lines of traps placed at right angles to the current than in those parallel to it. Sizer comments that frictional forces on these bottoms greatly reduce current velocities at the level of the drill and may explain the lack of response. Stauber emphasizes the fact that on most oyster planted grounds the roughness of the bottom probably not only greatly reduces current velocities in the immediate microenvironment of the drill but also establishes local eddy currents which tend to promote aberrations in the response of the drill to the main overlying flow of the current. In a migration experiment on smooth sandy intertidal bottom he did find that out of a total of 164 drills recovered, 72% moved in the directions of the tidal flow, and 28% at right angles to the flow. Galtsoff et al. (1937) in further migration studies in Delaware Bay noted that oyster drills moved against the current to some extent toward drill traps over bottom devoid of oysters. Haskin (1937) in careful studies in an intertidal area in Delaware Bay found that the oyster drill exhibited great variability in rate and direction of movement over the bottom, and that direction of current flow and the position of oysters were the primary factors in determining the orientation of the drills

Tidal currents are also of considerable importance in the dispersal of young drills on floating objects, and in the transportation of chemical substances emitted by prey which aid the drill in the location of food.

Gravity

Federighi (1931c) in studies at Hampton Roads, Virginia, discovered that drills exhibit a pronounced negative geotaxis, creeping upward when the temperature of the water rises above 10°C. This response is especially evident during the breeding season. Federighi noted that the response persists in a dark room, and with the eyes removed; and is not dependent on oxygen. Further, on a vertical glass plate turned as a wheel on a hub the snail always turns so that the siphon points up. Galtsoff et al. (1937), Stauber (1943), and the writer have noticed that during the colder months of the winter drills return to the bottom and many assume positions just below the surface. This appears to represent a reversal of the response to gravity at low temperatures.

Light

Federighi (1931c) under his experimental conditions was not able to detect a response to light on the part of the drill. However Sizer (1936) under different circumstances determined that direct sunlight has a pronounced effect on them. He placed drills in the center of the bottom of a covered aquarium in direct sunlight. The walls of the container were painted black except for the lower half of one side. The temperature of the water was kept below 25°C. About 70% of the drills in the experiments moved away from the source of light, and the remainder exhibited no definite response. When the intensity of the light was reduced on a cloudy day and by placing a sheet of white paper in front of the tank, the drills continued to move away from the light source. As the intensity of the light was reduced to that of a 75 watt bulb the drills moved slightly toward the light. At weaker intensities the tendency of the drills to orient toward the light increased; in dim light the phototactic response was lost completely.

Sizer saw a similar response in the field: in shallow water up to four feet deep on sunny days Urosalpinx were found attached to the under side of oyster clusters. On cloudy days they occurred on the top of the clusters, and in turbid waters of Delaware Bay at depths of about 15 feet more drills were taken on the top of the clusters than on the lower surfaces. Stauber (1943) rarely found drills on the most exposed portions of the reef on the Delaware Bay Cape Shore, and believes that the value of the negative response of drills to all but low intensities of light is more apparent here than in the deeper water of Delaware Bay.

Cole (1942) confirmed these observations. In a study of the rate of movement of the English Urosalpinx in a wooden trough filled with sea water he observed that until the intensity of sunlight was reduced by shading, the drills tended to move out of the sun into the shady side of the trough in bright weather. The writer has repeatedly observed this response in drills confined in aquaria over long periods of time. The influence of the length of day on the activities of Urosalpinx has not been investigated.

Ectocrines

The aqueous environment in which the oyster drill lives is enriched by an ecologically important but relatively unexplored complex of chemical substances which are released by living organisms or result from decay after death. Lucas

77

(1947) terms these external metabolites "ectocrines", and Bullock (1953) extends Lucas' concept to include the effective chemical signals involved in carnivorous sea star and gastropod predator-prey relationships. During the present century it has been demonstrated that a positive chemotactic response to the external metabolites of certain prey is also important in guiding Urosalpinx to food. It is instructive to review the evolution of this concept as it applies to U. cinerea.

That smaller oysters are preferentially selected by Urosalpinx has been common knowledge for some time. Pope (1910-11) and a number of workers after him believed that oyster drills select the thinner shelled bivalves, but were unable to explain how these bivalves are detected. Federighi (1931c) noted in the laboratory that drills will move toward living oysters in preference to oyster meats. T. C. Nelson (Haskin, 1950) observed that drills when feeding on a population of oysters of varied sizes do not always drill the smallest oysters. Many years later Cole (1942) in a similar observation in England noticed that drills show a distinct preference for spat of thumbnail size, and do not attack very small spat until all the larger spat are destroyed. And Stauber also (1943) reports a case in which five recently drilled oysters over 3 cm. in length were found among 85 young live oysters 0.5 mm. in diameter in Delaware Bay.

Haskin (1940, 1950) in laboratory studies performed during the summers of 1935 and 1936 in Barnegat Bay, New Jersey, provided considerable information on the degree of attraction to adult drills of substances released by living oysters of varying ages. He utilized a system of tanks in which sea water from an overhead reservoir, refilled at each high tide, flowed into compartments of a double tank. Oysters of different ages were placed in these chambers and overflow water was piped to opposite corners of a rectangular flat dish in the center of which experimental drills were grouped. Since Federighi (1929) demonstrated that drills move against a current of water, Haskin observed special precautions (not described) to eliminate the effect of current so that drills would orient only to substances from the oysters in the overflow water. In an extended series of observations he demonstrated conclusively that substances released by oysters and carried in sea water definitely attract oyster drills. He further demonstrated that water from one year old oysters is more attractive to Urosalpinx than that from two and three year old oysters; and water from two year old oysters is more attractive than that from three year olds. Oysters three years old or older are not distinguished from each other

In the early summer of 1937 Haskin (1950) extended his laboratory studies to field investigations on a sandy mud intertidal flat in Delaware Bay. Groups of oysters were placed in opposite corners of a 10 foot square. Marked drills were planted in the center and their positions were recorded during the following low tide.

These results confirmed his laboratory findings. He also discovered that older oysters collected in the brackish water of Maurice River Cove where growth is extremely slow, when placed in the experimental square with younger oysters, proved conspicuously more attractive to drills than the younger oysters. He explains that in the salty water on the flats of Delaware Bay growing conditions are more favorable than in the Maurice River Cove, and Cove oysters when transplanted to the flats were stimulated to rapid growth and a high rate of metabolism, a state characteristic of young oysters; Urosalpinx reacted to the older oysters as though they were very young oysters. These studies suggest that the metabolic rate of an oyster is more important than age in determining the degree of attractiveness of substances released by oysters. Haskin concludes that chemical attraction plays a dominant role in food selection by Urosalpinx.

Sizer (1936) demonstrated experimentally that drills are also attracted to oysters in deep water. He placed drills at one end and oysters at the opposite end of a cage and lowered it onto the bottom of Delaware Bay. During 24 hours most of the drills moved onto the oysters. He states that the drill attracting substance is probably a product of active metabolism which may be present in excretions and increases in concentration toward the oyster. The fact that a watery suspension of macerated oyster meats or excised meats from freshly opened oysters were inadequate in bringing about orientation of the drill, led him to state that the attracting substance is elaborated by the living oyster and does not occur to any appreciable extent in dead tissues.

In large scale drill trapping operations in Delaware Bay Stauber (1943) chose young oysters from the brackish waters of tributary tidal rivers for use as bait because of low cost and ease of handling, and inadvertently selected bait which competed in attractiveness with oysters of all ages on the Delaware Bay oyster grounds. In pilot tests two to three month old spat from the intertidal Cape Shore flats was also shown superior in attraction to older oysters, but did not prove practical bait on other grounds.

In both field and laboratory experiments Haskin found that three times as many drills were attracted to summer old oysters as to Mytilus edulis. The size and age of the mussels were not recorded. Since drills respond differentially to substances given off by oysters of different ages, they may respond in a similar manner to other food animals. The reported preference of drills for prey species other than oysters in some regions (Pope, 1910-11; Federighi, 1931c; Engle, 1935-36; Galtsoff et al., 1937; Cole, 1942; Stauber, 1943) may indicate a more pronounced attraction to the ectocrines of these species than to those of the oyster at the time of predation. Cole (1942) in explaining the lack of attractiveness of mussels to drills in the River Blackwater, England, where mussels are relatively scarce, suggests that Urosalpinx may evolve a predilection for those prey which occur commonly in its immediate habitat.

Cole (1942) could not establish that the drill is able to detect the presence of food and move toward it. He placed English Urosalpinx in the center of a wooden trough painted inside with pitch and filled with sea water. Young oysters or barnacles were set at varying distances from the center toward one end of the trough. During the experiments temperature ranged from 13 to 25°C, both bright sunlight and shade were employed, and the water in the trough was not moving. Under these conditions the drills moved at random, even when food was as close as 35 cm. Two conditions of the experiment may explain the lack of response of the drills to living food: the use of still water in which ectocrines diffused slowly and possibly irregularly, and a possible masking of the ectocrines by chemicals released by the pitch.

Stauber (1943) has noted that confinement of prey in aquaria for extended periods seems to reduce their activity and apparently their attractiveness to drills.

Galtsoff in recent experiments (pers. com.) has contributed further information on the detection of and movement of drills toward food. During the summer of 1954 he carried out 20 experiments in which 50 drills were placed in a large shallow water tank about 16 feet long and 8 feet wide in which the water moved parallel to the long axis of the tank at the rate of 1/2 cm./ sec. In a second set of experiments 20 drills were placed in a wooden trough 10 inches wide and 16 feet long. The water in this moved at the rate of 2 cm./ sec. A vertical partition divided one end of each trough into two parts. Adult oysters were placed on one side of the partition and seed oysters on the other. Drills were placed in each trough at the end opposite the oysters where they received water flowing past the oysters. The results were consistent in showing that approximately 50% of the drills began creeping toward the food. The remainder scattered about, sometimes climbing on the walls of the troughs, and remained inactive. The path of the active drills was a rather irregular spiral with many turns and circles. When approaching the oysters, the majority of the drills oriented themselves toward the seed oysters. The rate of crawling was inconstant and crawling was frequently interrupted by periods of inactivity. The rate of crawling while the drills moved in a straight line was about 1 cm. in 35 seconds. It is significant in terms of Cole's negative results that when the water was shut off in Galtsoff's tanks all the drills scattered in different directions and none reached the food within 48 hours

As the drill approaches a living oyster to which it has been attracted Stauber believes it may confirm the immediate presence of its prey by creeping to the excurrent stream of the oyster and/or by the shell movements of the bivalve. Neither possibility has been suggested before and no data are available to evaluate their significance.

It is evident that the inter- and intraspecies food preferences of U.cinerea of all ages should be reinvestigated on the basis of the growth rate, age, and geographic location of isolated individuals as well as of aggregates of naturally associated prey species. The response of very young drills to the ectocrines of newly set bivalves especially merits investigations.

Interrelations of Environmental Factors

An analysis of individual environmental factors which collectively regulate the behavior of the drill has merit as an aid to study and to interpretation. However, Galtsoff et al. (1937) and Stauber (1943), among others, are quick to point out that the behavior of the drill is determined by the combined and antagonistic effects of the environmental factors and cannot be attributed to a single cause. This emphasis has special relevance to this synthesis.

Biological Races

The existence of "physiological species" in drill populations of widely separated geographical areas has been indicated by Stauber (1950). He refers to groups of drills which morphologically belong to the taxonomic species Urosalpinx cinerea but which functionally differ in their response to such ecological factors as temperature. Until populations of "physiological species" are shown in fact to be reproductively isolated from each other, it may be better to refer to them as "physiological races". Stauber has drawn his evidence from the existence of these races from an analysis of data from Hampton Roads, Virginia (Federighi, 1931c), Delaware Bay, New Jersey (Stauber, 1943), and Narragansett Bay, Rhode Island (Gibbs, in Stauber, 1950), on temperatures which initiate locomotory movement, drilling, and oviposition in the oyster drill in these three regions within its range along the eastern coast of North America. The field data which Stauber presents (Table 13), though not of comparable scope or detail, does appear to support his hypothesis that physiological races exist in Urosalpinx.

Stauber (1950) strongly emphasizes the need for carefully controlled laboratory studies of physiological races. Loosanoff and Davis (1950-51) in a laboratory study, designed to determine the variation in the initiation of activities of oyster drills from 6 different geographic regions within a series of temperatures spaced at 3°C intervals, have made an excellent beginning in their attempts to determine whether these geographically isolated populations belong to different physiological races. They emphasize the fact that their unpublished report is a

81

TABLE 13. Relation of Water Temperature, ^{o}C, to Initiation and Cessation of Various Activities of <u>Urosalpinx</u> <u>cinerea</u> in Different Geographic Regions as Reported by a Number of Investigators

Movement Starts	Movement Ceases	Feeding and Drilling Starts	Feeding and Drilling Ceases	Oviposition Starts	Region	Source
-	-	11-12	11-12	12-13	England	Cole, 1942
		15	12.8	20	Canada	Adams, 1947
				11.1	Rhode Island	Gibbs (Stauber, 1943)
				10.9-11.2	Long Island Sound	Loosanoff, 1953
		11.6			Long Island Sound	Engle, 1953
4.5	2	9.5	9.5	13.9	Delaware Bay	Galtsoff et al., 1937
5	5 or below	10	10	15	Delaware Bay	Stauber, 1943, 1950
		12	8		Delaware Bay	Engle, 1953
		11.6			Maryland	Engle, 1953
10	10	15	15	above 20	Virginia	Federighi, 1931c
		6.5			Virginia	Andrews & McHugh, pers. com.
10	10	15	15	above 20	North Carolina	Federighi, 1931c
			7.6		North Carolina	Engle, 1953, of Federighi, 1931c

gions Responding by Turning Over, Feeding, and Oviposition to Different Temperatures (After Loosanoff & Davis, 1950-51)

Response	°C	Rhode Island	Long Island Sound	New Jersey	Virginia	North Carolina	South Carolina
TURNING OVER:	3	53.4	49.4	17.4	12.0	2.7	8.0
	6	53.4	64.0	22.7	25.3	54.7	44.0
	9	65.4	78.6	52.0	30.7	85.3	49.4
	12	77.4	84.0	62.7	45.4	81.4	65.4
	15	94.6	93.4	80.0	61.4	82.7	73.4
Average:		68.8	73.9	46.9	34.9	61.4	48.0
FEEDING:	9	14.0	10.0	4.1	2.2	4.1	4.1
	12	58.2	48.6	29.7	17.2	27.9	18.2
	15	69.0	76.0	48.0	39.5	44.0	42.0
	18	83.0	63.7	39.6	45.5	32.7	50.0
Average:		55.8	49.6	30.4	26.1	27.2	28.6
OVI-POSITION:	12	1.1	0.0	0.0	0.0	0.0	0.0
	15	12.4	5.4	3.5	0.0	1.3	0.0
	18	61.7	18.2	21.2	12.1	2.2	7.1
Average:		25.1	7.9	8.2	4.0	1.2	2.4
Total Average:		149.8	131.3	85.5	65.0	89.7	79.0
Average of Total Averages:		49.9	43.8	28.5	21.7	29.9	26.3

83

preliminary one and that their results may be modified in the future; however, because of the significance of their studies, their results have been included in this review, with the hope that more research of this nature will be stimulated.

Shipments of drills arrived in Milford, Connecticut, in the fall of 1950 from Rhode Island (courtesy of H. N. Gibbs), Long Island Sound (Milford Staff), New Jersey (H. H. Haskin), Virginia (J. D. Andrews), North Carolina (A. F. Chestnut), and South Carolina (G. R. Lunz). Drills from each region were dyed a different color for purposes of identification and then were maintained under identical conditions until January of the same year, when the laboratory experiments were begun. The following aspects of drill behavior were studied: temperature at which (1) drills were able to righten themselves from an inverted position, (2) feeding began, and (3) oviposition commenced. Groups of 25 or 50 drills were used in each experiment and almost all experiments were repeated three times; the same drills were usually utilized only once. After acclimatization for an hour at each experimental temperature, a mixture of drills from all geographic regions was placed in a large tray with sea water maintained at the experimental temperature. A summary of the results is presented in Table 14.

Turning over. Drills were placed at random in straight lines in the experimental tray. As soon as they turned over onto the foot they were removed. At a temperature of 0°C a few drills, principally from Rhode Island and Long Island Sound (and one from Virginia), extended the foot slightly but did not turn over. At this temperature no drill was able to attach to the bottom even if placed right side up. At 3°C approximately 50% of each of the Rhode Island and Long Island Sound drills rightened themselves within 90 minutes, whereas only a few from the other areas were able to turn over. The higher degree of activity of the North Carolina drills between 6 and 15°C than drills from South Carolina and Virginia seems to set them apart from the latter. There appears to be no distinction between the New Jersey and South Carolina drills at 9°C since at this temperature both exhibited 50% rightening in 90 minutes, while at temperatures of 12 and 15°C the Virginia drills appear to be the slowest to respond.

Feeding. A few drills attempted to feed in the Rhode Island, Long Island, New Jersey, and Virginia groups at 6°C; but no feeding was attempted by the North and South Carolina drills. At 12°C a significant difference was observed between the two northern groups and all the others, since approximately 50% of each of the Rhode Island and Long Island groups of drills and less than 30% of any other group fed. No appreciable differences were observed among the four southern groups.

Oviposition. Unfortunately these experiments were conducted either not long enough or at insufficiently high temperatures to obtain 50% egg deposition in any but the Rhode Island drills. No oviposition was recorded for any of the groups at 9°C even though the experiment was extended for 14 days. It may be significant, however, that the Rhode Island drills not only deposited egg cases at a lower temperature (12°C) than any other group but also that 61% of this group deposited cases at 18°C, whereas oviposition in the nearest two groups, Long Island Sound and New Jersey, only reached about 20%. The fact that 21.2% of the New Jersey drills oviposited at 18°C, a temperature considerably higher than that of the southern groups, may further separate the New Jersey drill population from the latter. Loosanoff and Davis explain further that the somewhat inconsistent results in the oviposition rates of the Long Island Sound drills may be explained, in part, by the possible preponderance of males in the samples. They also point out an interesting inconsistency when relating their laboratory findings to field observations where they (Loosanoff, 1953) have found recently deposited drill egg cases in 30 feet of water at a temperature of only 10.9°C, or approximately 1° lower than the lowest temperature given in Table 14, and at which in the laboratory experiments the Long Island Sound drills laid no eggs.

Loosanoff and Davis conclude from the averages presented in Table 14 that Rhode Island and Long Island Sound drills appear to differ from the southern groups in the higher percentage of the northern drills that commence turning over, feeding, and ovipositing at lower temperatures than do the southern drills.

The writer has included additional field data relating to physiological races of Urosalpinx in Table 13 for comparison with Stauber's (1950) data and that of Loosanoff and Davis (1950-51) in Table 14. Table 13 discloses a number of apparent inconsistencies which may best be explained on the basis of incomparable accuracy of many of the observations and/or ecological differences within the specific habitats in which the drills were observed or collected. It is significant that Loosanoff and Davis obtained movement of drills in all areas at lower temperatures than those reported in Table 13, and especially for Virginia and North Carolina. Since Stauber's figure of 5°C for movement of drills in Delaware Bay is based on movement into traps his figure would be higher than that obtained in the close laboratory observations.

Federighi (1931c) observed feeding and drilling at 15°C in Virginia and North Carolina, while Engle (1953) reports that it occurs as low as 7.6°C in North Carolina, and Andrews and McHugh (pers. com.) recorded it at an average temperature of 6.5°C in Virginia. Loosanoff and Davis' data indicate that feeding and drilling may commence at 6°C, a figure considerably lower than that given in the field data in Table 13. Again where data were collected by means of drill traps a lag would be expected between drills in situ on the bottom and those which move into traps to feed.

85

Gibbs' figure of 11.1°C for onset of oviposition in Rhode Island drills coincides with Loosanoff and Davis' figure of 12°C, particularly since the latter worked at temperatures spaced at 3° intervals. Galtsoff et al.'s figure of 13.9°C for the initiation of oviposition in New Jersey is somewhat lower than Stauber's figure of 15°C, but compares favorably with a spawning temperature somewhere between 12 and 15°C in Loosanoff and Davis' data. Adams' figure of 20°C for oviposition in Canada is noticeably higher than the relatively close figure of 11.1°C for Rhode Island, 12-13°C for England, and 10.9-11.2°C for Long Island Sound, but close to Federighi's figures of 20°C for Virginia and North Carolina. The similarity of response of drills in North Carolina, Virginia, and Canada may reflect the sheltered, shallow, comparatively warm waters in which the Canadian drills normally reproduce. The figures for initiation of oviposition in England, 12-13°C; Rhode Island, 11.1°C; and Long Island Sound, 10.9-11.2°C, are also close. The writer concurs with Stauber's (pers. com.) suggestion that this similarity indicates that drills originating in the New England area are probably the ones which survived after transportation to England on oysters. It has been shown in the section on distribution in this review that large quantities of live oysters have been shipped to Great Britain from the New York area over long periods of time. Similarly it has been shown that many of the live oysters shipped to the west coast of North America came from the New York area, and thus probably carried drills from these cool waters to those of the west coast where they have been able to reproduce. It remains to be demonstrated on a controlled experimental basis to what degree drills from warm geographic regions will continue to reproduce when transplanted to cooler regions

Some of the data in Table 13, though limited, further suggests that drills may start moving in the spring at a temperature, 4.5°C (Galtsoff et al., 1937), higher than that, -1 to 2, at which they cease movement in the fall; and may start drilling and feeding in the spring at a higher temperature, 15°C (Adams, 1947), than that, 12.8°C, at which they cease in the fall. The temperatures shown in Table 13 and dates in Table 4 at which oviposition commences in various regions do not necessarily corroborate Stauber's observation that even in widely separated areas and in the presence of considerable differences in water temperature, spawning of the drill begins within a relatively short period in the spring.

The low temperatures at which Cole (1942) obtained maximum locomotion rates in British drills, as compared to the higher temperatures at which Federighi (1931c) observed maximum rates in the southern United States, if confirmed, might introduce further support for the existence of physiological races. The relation of minimum salinity tolerances of drills to drill races is not known. Nor is information available on the function of acclimatization in the evolution of physiological races segregated on the basis of response to temperature (Stauber, 1950).

That at least some of these races adjust to transplantation to different coast lines of the world is amply demonstrated by the success of Urosalpinx introduced in England and on the west coast of America

In summary it may be stated that Stauber's hypothesis of the existence of physiological races receives added support from the laboratory experiments of Loosanoff and Davis and from some of the additional information in Table 13, though it is evident that with refinement of experimental technics and more careful observations on the quantity of heat necessary to initiate physiological activities the lower thermal limits at which these physiological activities begin in drills from different regions may be shifted. This should not alter Stauber's evidence for the existence of physiological races. On the other hand, Jenner (pers. com.) points out that the evidence for the occurrence of physiological races as presented by Stauber (1950) is based on the assumption that temperature is the controlling factor; if it should turn out that a factor such as day length is critically important, then Stauber's conclusions would not be supported.

This portion of this review has indicated a number of suggestions for future research on physiological races of the drill. Since in the laboratory, organisms, particularly those from te m p-e-r a te regions, exposed to variable temperatures frequently show accelerated development as compared with those held at a constant temperature of the same mean value, other conditions remaining the same (Allee et al., 1949), it is important that activities of Urosalpinx under both sets of conditions be compared. The difference in the onset of oviposition obtained by Loosanoff and Davis (1950-51) may be explained in part by the constant thermal conditions in the laboratory and the fluctuating thermal conditions in the field. In addition comparative studies on the rate of oviposition in drills will not be dependable until the ratio of females to males used is accurately determined. This points to the need for a method of reliable sex determination in living drills; or less desirably, as Jenner (pers. com.) has suggested, that the sex of the experimental animals be determined by anatomical examination at the conclusion of the experiments. Finally, since oviposition in Urosalpinx may occur only after the summation of a specific quantity of heat energy, it would seem necessary to design experiments with this in mind. On the other hand length of day may be an important factor in influencing the reproductive cycle of the drill, so the study of oviposition under artificially manipulated photoperiods should be undertaken (Jenner, 1954). No investigations of this nature have been reported for the oyster drill.

The existence of morphological races has also been reported. Walter (1910) in a comprehensive study of the relative morphological variability expressed by U. cinerea living in native habitats and when introduced to new environments, recorded the ratio of the maximum dimension of the shell aperture to the total shell height in a total of 50, 424 drills collected in San Francisco Bay, California; Princes Bay, Staten Island, New York; Norwalk Harbor, Connecticut; Cold Spring Harbor,

New York; and Woods Hole, Massachusetts, during 1898-1908. He concludes that so far as his statistical method reveals, it is extremely doubtful whether or not Urosalpinx, such as those inadvertently transported from Staten Island to San Francisco Bay, exhibit greater variability of this ratio when introduced to new habitats. He notes that since a change in variability accompanies drills during growth it is practically impossible to collect homologous lots of individuals on which environmental modification may be accurately determined. In spite of these difficulties he reports that drills from different native habitats vary widely enough from each other to be easily distinguished. In particular he observed that drills from localities more exposed to the beat of the waves show greater variability than those from more protected places; and that drills in dense populations express less variability than those in sparse populations.

At least two, and possibly three, subspecies of U. cinerea are known to occur. Two are quite distinct, the small widely distributed form described as the typical U. cinerea, and a giant form, U. cinerea var. follyensis (Baker, 1951) from the eastern shore of Maryland and Virginia. The maximum height of U. c. follyensis collected by Henderson and Bartsch (1915) measured 51.5 mm., by Baker (1951), 51.2 mm., and by Galtsoff et al. (1937), 61 mm. (Table 7). Owen (1947) in a comparative cytological study of spermatogenesis in the male gonads of these two subspecies shows that the diploid chromosome number (32) is constant in both, but noticed that in all cases during Anaphase I of the typical U. cinerea one chromosome lags behind the others forming a distinct chromosomal bridge. This chromosomal behavior characteristic is not present in U. c. follyensis. Owen concludes that Henderson and Bartsch (1915) were correct in recognizing two separate races in the species U. cinerea.

Abbott (1954) suggests that Urosalpinx perrugata Conrad from the west coast of Florida, which is similar to U. cinerea, may be another subspecies of U. cinerea.

RELATION TO OTHER ANIMALS
Cannibalism

Some cannibalism occurs among U. cinerea of all ages in the presence of other food both in the field and in confinement. Pope (1910-11) reports that in one observation 100 newly hatched drills hatched in captivity were reduced to 36. T.C. Nelson (1922) doubts that such extreme cannibalism occurs in nature, since drills do not all hatch simultaneously and tend to scatter. Stauber (1943) also detected cannibalism among recently hatched drills in aquaria, but an amount less than the extremes reported by Pope. Pope (1910-11), Haskin (1935), Galtsoff et al. (1937), Stauber (1943), and F. B. Flower (1954, New Jersey Oyster Research Laboratory) report cannibalism among adult drills in both the laboratory and in the field.

88

Flower during a series of dredgings in lower Delaware Bay collected bottom material retained on a 1/4 x 3 inch mesh screen. In this material he counted 937 dead drills. Seventy-six of these had been drilled by other Urosalpinx and Eupleura. The figures of dead drills undoubtedly represent an accumulation over a long time.

Other Species of Drilling Gastropods

F. B. Flower (1954) in his examination of dead drills from Delaware Bay reported in the preceding section, made the startling observation that 100 of the 937 dead Urosalpinx and Eupleura had been perforated by a species of Polinices. By confining three Polinices duplicatus and 23 Urosalpinx cinerea in an aquarium he obtained similar destruction of Urosalpinx by Polinices. At the end of four months only one moon snail and one oyster drill were alive. The other two moon snails and 22 drills had been drilled and consumed by the moon snail(s). Predation of oyster drills by Polinices has not been reported before.

Cole (1942) and Orton and Lewis (1931) clearly demonstrate the progressive replacement of the English drill Ocenebra by Urosalpinx in the River Blackwater. The other English drill Nucella has also decreased markedly in recent years. Some of this replacement may be attributed to the higher tolerance of Urosalpinx to severe winters. How aggressively Urosalpinx has competed for food and other requirements is not known.

Sea Stars

Mead (1900) noted that sea stars, especially when young, are exceedingly voracious feeders, and prey upon a number of organisms including oyster drills. He suggested that Urosalpinx may be held in check to some extent by Asterias. Coe (1912) stated that A. forbesi is occasionally of actual benefit in that it devours U. cinerea, but does this only when oysters or other bivalves are not obtainable. This has been corroborated by Engle (1940, 1954), A. Flower (Glancy, pers. com.), and Loosanoff (pers. com.). Engle (pers. com.) believes that when sea stars and oyster drills exist together a considerable number of drills are consumed by the stars. That sea stars even when present in large concentrations in company with oyster drills have not eradicated the drills is shown in Long Island Sound (Glancy, pers. com.; Engle, pers. com.), and Shark River, New Jersey (Carriker, unpub.). Neither is it known whether Urosalpinx control the abundance of sea stars. A series of observations by the writer in Gardiners Bay, New York, during the summer of 1953, very strikingly demonstrated the almost total absence of sea stars and a phenomenal abundance of Urosalpinx in areas where temperature of the water did not rise above the tolerance limit of sea stars. It is recognized, of course, that this condition may represent a chance disproportion of these species without relation to possible predatory control, particularly of small sea stars by drills.

Bullock (1953) has reported upon predator recognition and escape
responses of some intertidal gastropods in the presence of carnivorous sea
stars on the west coast of the United States, and states that a number of gastropods exhibit specific patterns of escape. However, a muricid carnivorous snail,
Acanthina, which he studied displayed no such response. A similar study of the
predatory-prey relationships of Asterias and Urosalpinx should prove illuminating.

Fish

Goode (1884) was informed by a bayman in New York that in the years when
eels (Anguilla rostrata) were plentiful, the oyster drills were "kept down" because
the eels fed on their egg cases. This observation has never received confirmation.

Oystermen in Delaware Bay reported to Stauber (1943) that toadfish
(Opsanus tau) consume drills, so he examined the stomach contents of a number of
them. In every case but one he found that the drill shells therein contained hermit
crabs (Pagurus longicarpus) in the process of digestion. McDermott (1952) during
the summer of 1952 analyzed the stomach contents of 37 toadfish, also collected
in Delaware Bay, and encountered only the empty shells of two Urosalpinx. He
reports that small Crustacea constituted the bulk of organisms in these stomachs.
It would seem that these fish are not predators of the oyster drill.

Parasites

Stunkard and Shaw (1931) record the presence of the larval trematode
Cercaria sensifera in Urosalpinx in the Woods Hole, Massachusetts, area; and
later Cole (1942) found the same parasite in a single drill from the River Blackwater, England. Stauber (1941) has reported the presence of an ectoparasitic
polyclad turbellarian, Hoploplana inquilina thaisana, in the mantle cavity of drills
in Delaware Bay. In addition Cole (1942) found the digestive diverticulum of a
number of drills in the River Blackwater parasitized by an unknown arthropod,
probably an isopod. No mortality has been associated with the presence of these
parasites. There are no reports on the possible role of these parasites in the
biological control of the drill.

MIGRATION
Horizontal

Since U. cinerea lacks a free swimming larval stage, it depends entirely
on its own slow rate of creeping or upon transportation by human or other agencies
for its distribution. Much interest has developed during the last 20 years, particularly in oyster culture, on the rate and extent of migration of this species in the
field. The first to apply an experimental approach to this problem was Federighi
(1931c).

He released an unknown number of drills marked with pigment around a stake on oyster ground in deep water in Hampton Roads, Virginia. Concentric circles were marked at 25 foot intervals from the stake and 4 to 6 stations on each of these circles were sampled for marked drills with tongs at weekly intervals. In a period of 2-4 weeks the following marked drills were recovered: at the stake, 2; at 25 feet, 9; 50 feet, 8; 75 feet, 2; 100 feet, 6; and 150-200 feet, 1. These figures indicate that the majority of the drills moved at a possible maximum rate of 7 feet per day or less; the fastest drill, 10 feet per day. The recovery of only 28, or 5%, of the marked drills is explained by Federighi as due to the inefficient method of sampling and the possible loss of the identifying pigment.

Federighi considered the Hampton Roads studies inconclusive so performed further experiments in Beaufort, North Carolina, on oyster beds exposed or slightly under water at low tide. Drills tagged with a numbered celluloid tag fastened to the outer lip of the shell with a fine silver wire were planted around a stake and the movements of each drill were followed daily at low water. In every case, even after one month, he found that the tagged drills had not moved over 10 to 15 feet from the stake, an approximate rate of 0.4 to 0.5 foot per day. He believes that this low rate was not due to the presence of ample food, because in one experiment drills placed on firm bottom about 20 feet from an oyster bed in no case moved to it.

Federighi's data do not show that the results obtained in Beaufort corroborate those in Hampton Roads as he suggests, since in Beaufort drills moved no more than 15 feet per month and in Hampton Roads a number migrated a possible 100 feet in the same time. His data suggest that the Beaufort drills, as indicated by their low rate of movement, were impeded by the celluloid tag.

Federighi concludes that the drill does not migrate extensively and that its distribution has been accomplished principally by man in the culture of oysters.

Sizer (1936) in field experiments in Delaware Bay deposited 711 marked drills in the center of a 20 foot square surrounded by 28 drill traps on moderately soft bottom in water about 15 feet deep. In four days when traps were tended he found that 23% of the drills had travelled a distance of approximately 10 feet to the bait, or about 2.5 feet per day. He repeated the experiment on hard bottom with similar results but obtained a lower recovery of drills. Unfortunately traps were not tended frequently enough to show the time of arrival of the marked drills at the bait, and thus the actual rate of travel. Sizer states that the tides had no apparent effect upon the direction of migration.

During 1935-36, over a period of several summer months, Galtsoff et al. (1937) carried out some migration studies in Delaware Bay, New Jersey, in which they marked drills by dipping the spire into quick drying colored lacquer which persists for several months. These drills were released in the center of large areas on a variety of firm grounds devoid of oysters and shells and at depths varying from 15 feet to low water mark. Wire cages baited with young oysters or filled with shell were placed at successive intervals from the drills and examined daily. In every case more marked drills migrated to the oyster baited cages than to those filled with shell. Engle (1935-36) who has described some of these results in more detail reports a maximum rate of 75 feet per day but shows that daily migratory rates in excess of 18 to 24 feet per day were unusual. Twenty-nine to 67% of the marked drills were recovered. These authors conclude that in search for food U. cinerea is able to travel a considerable distance, and that high drill densities may be correlated with the presence of an abundant food supply. It should be noted that for the most part Federighi carried out his experiments on bottom covered with oysters whereas Galtsoff et al. used barren bottom and baited drill traps These authors stressed the maximum distances covered by the most active drills.

In a total of 6 preliminary field experiments on the intertidal flats of Delaware Bay, Haskin (1937) studied the movements of Urosalpinx in relation to the factors controlling the direction of migration during the period June 15 to July 13, 1937. The bottom of these flats is composed of a loose mixture of sand and mud and is irregular in contour, so that at low tide many bars are exposed while some sloughs contain as much as a foot of water. During the studies water temperatures taken daily usually at high water ranged from 18.0 to 33.7°C. Salinities in the same period varied from 18.5 to 23.6 o/oo. As a check on the direction of net current flow, floats, weighted with small rocks so the weights remained on the bottom yet required only a very slight current to move them along, were followed from one low tide to the next. A total of 1,309 drills marked with quick drying pigmented shellac were released on bottom relatively free of food organisms, principally during low water, and at stakes oriented with respect to reefs and bags of living oysters and direction of current flow. Observation of the position of the drills was made mostly on succeeding low water periods. In general percentage recovery of released drills decreased with time, ranging from 76% in observations extending a little over one day to 11% in an observation continuing for 7 days. Haskin found that U. cinerea exhibited great variability in rate and direction of movement over the flats, and that direction of current flow and the position of oysters were the primary factors in determining the orientation of the drills. The apparent variability in drill responses among the different groups resulted from variations in the relative strengths of chemotactic and rheotactic stimuli in the areas of the different stakes. Some drills averaged as much as 4.4 feet per hour for the period of 5 hours after liberation (in one case a drill moved 6 feet in less than 1/2 hour after liberation), but a rapid

92

decrease in the rate of locomotion occurred shortly after this. In all cases the average rate of movement of the fastest moving individuals steadily declined toward the group level as time of creeping from the time of liberation increased. This is an important consideration which other investigators have overlooked. Neglecting burrowers and obvious laggards Haskin obtained an average rate of movement of 21 feet per day for the bulk of the liberated drills in a period ranging from 2 to 7 days in the 6 experiments; and in the same period an average of 32 feet per day for the few farthest ranging individuals. Maximum distances covered by the latter in individual experiments in periods of 2 to 3 days averaged 58 feet per day. The maximum distance covered by any single drill during the first 24 hours was 92 feet, but Haskin found that these far ranging drills were transported in large part by hermit crabs (see later, this section). He also noticed that a large proportion of drills burrowed into the bottom and remained completely covered for several days. This would account in part for the low recovery of released drills.

The most recent studies on drill migration have been performed by Stauber (1943) who also emphasized the distance covered by the majority of a migrating population. He, like Haskin, believes this is important since transportation of drills by other animals may explain some of the extreme values reported by Federighi and by Engle. Stauber's first experiments were performed on relatively clean sandy mud bottom just below low water mark in Delaware Bay. During the five trials reported the water temperature ranged from 23 to 25°C. Drills dipped in pigment were released in the center of a square mostly during high water, and the extent of their movements was checked some 14 to 20 hours later at low water. Oyster baited drill traps were placed at the corners of the square. Eighty-two percent of the 563 drills used in the observations were recovered, and the majority were found buried just beneath the sand attached principally to the tubes of a burrowing worm. In all, 91% of the recovered drills moved less than 10 feet in the 14 to 20 hour period.

From September to November Stauber utilized the same quadrat for 6 more trials in which marked drills were left undisturbed for a period of one week. The temperature of the water continued to drop during the experiments. Of a total of 445 drills placed in the square, 14% failed to move 10 feet and 47% crawled to and were recovered almost exclusively in the drill traps.

The following fall L. Jeffries working with Stauber carried out a drill migration study in deeper water on bottom ordinarily utilized for oyster farming but free of oysters at the time of the experiment. An area roughly 800 feet square was staked and 1,979 marked drills were released in the center. Three hundred and eighty-one oyster baited drill traps, which were tended weekly for a period of two months, were stationed along the boundaries of the square. During the first

93

month the temperature of the water dropped gradually from 24.2 to 16.4°C, and by the end of the second month, to 10°C. The tidal currents flow north and south over this bottom. In the two months of trapping only 4% or 81 of the marked drills were recovered. Ten of these were captured between the 7th and the 11th days, suggesting (if the drills travelled in straight lines....which is not generally the case) that they moved roughly 400 feet in 7-11 days, or 36 to 57 feet per day. The maximum catches of drills occurred 19 to 26 days after the marked drills were released, connoting a migration rate for the majority of the recovered drills of 15 to 21 feet per day. Although 63% of the total number of drills recovered moved southward toward oyster plantings, 7.4% moved west toward bottom with but a few oysters and at right angles to the flow of the currents, and 9.9% moved east and 19.7% moved north both to areas devoid of oysters.

Needler (1941) states in summary of his observations which he does not describe that drills in Canada move only an average of 10 to 15 feet in a month, that they show no definite horizontal migrations, and that they spread quickly only if carried.

In his drill trapping operations Stauber obtained much information on the spottiness of drill densities over large areas of the bottom in Delaware Bay. This was particularly evident on adjacent grounds on which drill food supplies differed strikingly over long periods of time, and confirmed Federighi's earlier observations on similar irregularities in the distribution of the drill. One of Stauber's many examples will suffice: In 1936 he trapped three contiguous sections of bottom: the first held small oysters planted two months earlier; the center, old oysters planted four years earlier; and the third, oysters planted two months earlier but slightly larger than those in the first section. Traps, baited with freshly collected brackish water oysters, were tended weekly from July to September, and yielded insignificant differences among the three sections. On the basis of Haskin's (1950) results it would be postulated that drills on the large oysters would tend to migrate to the younger oysters. This apparently did not occur. In South Carolina Lunz (pers. com.) also finds an extremely spotty distribution of drills; some oyster beds are inhabited by drills while nearby grounds with apparently identical conditions are not infested. Newcombe and Menzel (1945) also report much variation in the concentration of drills from ground to ground.

Stauber has summarized additional information which tends to minimize the extent of migration by Urosalpinx. He found in drill dredging operations that the proportion of drilled Urosalpinx shells increased with the increased degree of barrenness of bottoms. He observes that it is difficult to reconcile such cannibalism with the maximum migratory rate reported by Galtsoff et al. (1937) and the relative proximity of oysters on adjacent bottoms.

As a result of his studies Stauber concurs with Federighi that drill migration is probably less important in influencing densities of drills on oyster bottoms than other factors. Cole (1942) from his observations on the rate of movement of drills in a trough and a study of the results of Federighi and of Galtsoff et al., concludes that it seems unlikely from the limited evidence available that migration is of more than local significance in the distribution of the species. Engle (1953) from many years of experience also concludes that the drill does not migrate very rapidly.

The data collected by various authors on migration in U. cinerea are summarized in Table 15. Inspection of this reveals that drills living among oysters and other food organisms move very little (Federighi 1 & 2) and that drills tagged with celluloid tags move scarcely at all (Federighi 2). The most reliable data available indicates that the majority of drills on oyster bottom devoid of oysters, or on clean firm bottom, tend to move at an average rate of 15 to 24 feet per day in the direction of food (Galtsoff et al., Haskin, Stauber 3). In this connection Adams (1947) describes an instance in Canada where drills crawled 75 feet across barren mud bottom to a dense planting of oyster spat, but unfortunately does not report the time involved and the consistency of the bottom. And Mistakidis (1951) states that in England drills favor well cultivated grounds and apparently migrate actively to these from surrounding muddy bottoms. He does not support his suggestion; it is more likely as other observers have suggested, that drills where possible avoid loose muddy bottom devoid of food. Table 15 further suggests that a few drills may be carried considerably distances (Galtsoff et al., Stauber 3, Haskin) --- see later, and that sand slows the movement of drills over the bottom (Stauber 1 & 2).

It has been mentioned that pronounced variability of drill densities even on adjacent grounds denotes a low rate of migration. Oysters and empty shells maintain a rich fauna (Norringa, 1951) and it is probable that many of these animals are utilized by the drill for food. Consequently a sizeable population of drills may be supported on oyster bottom after oysters grow to a size less vulnerable to drill attack. The writer suggests that the impetus for drills to emigrate from bottoms covered with shell or with older oysters may not be as strong as has been supposed, but that in the absence of food or in the proximity of an abundant supply of new food drills may exhibit pronounced movements. Stauber (1943) adds that drill locomotion is variable not only over unused bottom, but may be slowed considerably by patches of mud or by unevenness due to the presence of considerable shell.

Haskin (1937) and Stauber (1943) contribute the important suggestion that the transport of drills by other animals may account for the extreme distances covered by marked drills in migration studies. Federighi (1931c) was the first to

95

TABLE 15. Migration Rates of Urosalpinx cinerea in the Field (From Numerous Sources)

Investigator and Location	Type Bottom, Food, and Method of Capture	Temperature of Water	Method of Marking	Percent Drills Recovered	Migration Rate per Day Average	Migration Rate per Day Maximum
Federighi, 1931c:						
1. Hampton Roads, Va.	Among oysters, do p water; tongs.	(Summer)	Pigment	5%	1-7 ft.	10 ft.
2. Beaufort, N. C.	Among oysters; inter-tidal.	(Summer)	Tags	?	0.4	0.5
Sizer, 1936: Delaware Bay, N. J.	Soft oyster bottom; 15 ft. deep; oyster bait-ed traps.	(Summer)	Pigment	23	2.5	?
Galtsoff et al., 1937: Delaware Bay, N. J.	Clean b tt m devoid of shell and oysters; de p to shallow water; oyst-er baited traps.	(Summer)	Pigment	29-67	18-24	75
Haskin, 1937: Cape Shore, Delaware Bay, N. J.	Intertidal sandy mud; oyster reefs and oyster baited traps.	18.0-33.7°C	Pigment	11-76	21	58
Stauber, 1943:						
1. Cape Shore, Delaware Bay, N. J.	Intertidal sandy mud; oyster baited traps.	23-25	Pigment	82	less than 10	?
2. Cape Shore, Delaware Bay, N. J.	Intertidal sandy mud; oyster baited traps.	Sept.-Nov.	Pigment	61	less than 10	?
3. Delaware Bay, N.J.	Oyster bm devoid of oysters; oyster baited traps.	24-16	Pigment	4	15-21	57
Needler, 1941: Canada	?	(Summer)	?	?	0.3-0.5	?

report that drills may be carried long distances on crabs, especially the hermit crab, onto which they crawl to feed on the encrusting gastropods, crustaceans, and other sessile animals. Stauber (1943) reports similar observations.

Haskin(1937) in his intertidal migration studies had ample opportunity to observe the activities of the common hermit crabs, _Pagurus longicarpus_ and _P. pollicaris_, in transporting Urosalpinx. He notes that in regions where these crabs are so numerous as in Delaware Bay the role that they play in distribution of drills is probably a significant one. On two separate occasions he found five marked Urosalpinx attached to the shell of a Polinices occupied by a Pagurus pollicaris. In several instances drills were found clinging by their pedal surfaces to shells no bigger than their own inhabited by actively moving hermit crabs. In numerous cases hermit crabs occupying marked drill shells were found some distance from the site of release of the drills; the possible role of the crabs in killing the snails is not indicated. In one case a marked drill was found on the shell of a large hermit crab 12 feet from the site of liberation 15 minutes after release. In an extreme case Haskin recovered one marked drill, dead but still entire, being dragged over the bottom by four small hermit crabs 150 feet from the stake where it had been placed 10 hours before. J. R. Nelson (pers. com.) has picked up horsefoot crabs (Limulus polyphemus)carrying as many as 140 drills per animal. How common such unusual cases of phoresis are is not known.

Young drills may also be distributed by the current when they creep onto floating algae and debris (Federighi, 1931c; Stauber, 1943). J. R. Nelson (pers. com.) from years of observations in the coastal waters of the northeastern states writes that tidal currents are of great importance in the dispersal of young drills on inert objects that come to rest on the bottom at slack water and move when the current reaches sufficient velocity. He notes that areas like Fireplace, Gardiners Bay, Long Island, and Delaware Bay possess tidal currents of such velocity that such objects are doubtless transported several miles in a single tidal period. This would account for the rapid migration several miles upbay into normally drill free areas which he observed in Delaware Bay following the dry years of 1930 and 1931. T. C. Nelson (pers. com.) observed the transport of newly hatched Urosalpinx on Zostera floating in a tidal stream over a distance of at least 40 feet. He notes that young drills are strongly negatively geotactic hence would be transported on any material carried along the bottom.

On the basis of this review the writer tentatively supports the conclusions of Federighi and Stauber that the bulk of the drill population moves about to a rather limited degree, particularly over oyster bottoms; occasional exceptions may be explained on the basis of phoresis. Haskin's interesting observation on the decline of locomotory rate of fastest moving liberated drills to the group level

97

with time, emphasizes the danger of basing migratory rates on short periods particularly immediately after release of drills, and coincides with the limited rate of migration of drill populations reported by others

Additional detailed studies using marked drills on bottoms covered with varying concentrations of shell with oysters of varying ages, and with varying mixtures of shell and oysters, must be carried out before the complete story on migration can be presented. Such explorations should be carefully related to temperature, salinity, current velocity, turbidity, type of sediment, depth of water including intertidal areas, degree of concentration of shell and of oysters, available food organisms other than oysters, if on barren bottom the proximity of food organisms, size of the drills, the stage in the reproductive cycle of Urosalpinx, and time of release.

Spawning Migration

A number of observations demonstrate that a portion of the drills normally occupying subtidal bottoms in the winter migrate inshore onto favorable intertidal bottoms to spawn early in the summer. Gibbs (pers. com.) for many years has reported the absence of Urosalpinx in the intertidal zone in the winter and their presence there in the summer, and has supposed that this is explained by migration. Orton (1930) states that in England there is an inshore migration in the spring and the species spawns heavily in shallow water. Cole (1942) adds that although this spawning is evident in the intertidal zones of the River Blackwater, very large numbers also spawn at all depths. Mistakidis (1951) reports a similar inshore migration in the Rivers Crouch and Roach, England, during the spawning season.

Cole (1942) by simultaneous sampling of large numbers of drills in the intertidal zone and in deeper water in the River Blackwater in June and at monthly intervals thereafter in the intertidal zone alone, has shown that the first drills to come ashore in June are mainly four year old females and five and six year old males. The younger and older drills remain in deeper water. By July, three year old females increase enormously inshore and form the dominant group and thus appear to spawn about a month after the older drills. Older and one and two year old drills also appear but in lesser numbers. Among males the tendency is similar and as the season advances the younger age groups appear inshore in increasing numbers; but one and four year old males are never as numerous inshore as in deeper water. The largest numbers of drills were collected in June and July and by August many had already returned to deeper water. Cole's data also show that out of a total of 1,288 male and female drills collected in the River Blackwater in 1939 and 1941, 814 were females and only 474 were males. Whether it is a general rule that more females than males participate in the inshore migration cannot be

98

stated until more observations are made. No one has yet attempted to follow marked drills in the spawning migration.

Stauber (1943) has clearly demonstrated during his successive years of drill trapping in Delaware Bay that drills inhabiting deep water bottoms some distance from intertidal areas also exhibit a pronounced prespawning locomotory activity which he calls the "pre-egg laying activity". This may be similar to that expressed by drills migrating inshore. Using the number of drills captured per trap per week as an index of activity, Stauber obtained a marked rate of activity in May when the temperature of the water normally rises from 15 to 20°C in Delaware Bay. This peak of activity subsides conspicuously with the onset of oviposition even though the temperature of the water continues to rise for about 6 weeks more, and is not usually equalled again until July or August when young drills of the current season, which make up as much as 25 to 50% of the total catch, start moving into the traps

It should be emphasized that it is not known how far drills will travel during the inshore summer migration, nor has it been definitely proven that these migrations occur for spawning purposes alone. It is possible that the migrations express the negative response of drills to gravity at summer temperatures in much the same way that in deeper water they crawl onto elevated objects, and/or that drill population pressures add impetus to the migration. Finally, since some drills may bury in intertidal sediment over the winter, not all drills appearing intertidally in the summer may represent deep water migrating Urosalpinx.

"Sudden Appearance" of Drills on Oyster Grounds

The "sudden appearance" and unexplained origin of high concentrations of drills on oyster bottoms is reported by oyster growers from time to time. The writer agrees with Stauber (1943) that these phenomena appear to have a rather simple biological explanation.

Stauber (1943) demonstrated in Delaware Bay that even so-called vacant oyster bottom retains some drills and that grounds dredged free of market oysters in the winter may be infested with appreciable quantities of drills. In either case, as is generally the commercial practice, seed oysters are planted on these bottoms in the spring before the peak of oyster drill oviposition and supplement the food supply of the drills. During the summer the hatching drills also find added food on the newly planted seed oysters which regularly catch quantities of oyster spat, barnacles, and other sessile organisms. As a consequence the survival of the small snails is high, but the oyster grower who does not always carefully search for the pests and who uses only the ordinary oyster dredge as his sampling tool, generally overlooks them. By the end of the second summer most of the drills have grown to adult size and the drill population has been augmented by another spawning. It is at this time that oystermen usually notice wholesale destruction of young oysters on their grounds and erroneously attribute this to drills which have migrated from contiguous bottoms. Stauber believes that at times migration may play an important role, but that it is secondary to the considerations presented above.

A dramatic example of the "sudden appearance" of drills on newly planted oyster seed with disastrous consequences was reported to the writer recently by the J. & J. W. Elsworth Co. In the fall of 1951 an 80 acre plot in West Neck, Peconic Bay, Long Island, was drill dredged by means of the conventional oyster dredge with the 1/4 inch mesh bag. The ground was similarly drill dredged in the spring of 1952 and in August of the same year just before 4,000 bushels of New Jersey oyster spat, varying in diameter from 1/8 to 1/4 inch, were carefully trucked into Greenport and expeditiously planted on the prepared ground. During the course of the three successive drill dredgings a marked decrease in the number of drills was reported. But in two weeks most of the newly planted spat had been drilled, and in three weeks the entire planting was a commercial loss. The majority of drill holes in the destroyed spat were minute, and a minimum of two to four drills, mostly very small in size, were counted per cultch shell. Although it is reasonable to assume that the majority of the larger drills were removed from this ground during the successive drill dredgings, there is no assurance that many of the minute drills were not washed back onto the bottom through the 1/4 inch mesh bag of the dredge; the reported diminutive size of the drills and their perforations support this contention. In the short span of three weeks time during which the spat were destroyed it is inconceivable that, even though attracted by the young oysters, enough of these small drills could have migrated onto the bed, or been transported there on drifting debris or on the back of larger animals, to cause the damage reported. The very strong suggestion in this instance is that the spat were destroyed principally by young drills already on the ground at the time of planting.

F. B. Flower and associates, New Jersey Oyster Research Laboratory (pers com.), have been making observations in Delaware Bay in cooperation with the Claude Jefferies & Sons suction dredge the "Luther Bateman" which further confirm the fact that significant numbers of drills may be washed through the dredge screen. Material drawn off the bottom by the dredge is flushed onto two shaker screens, a one inch mesh and a 1/4 x 3 inch mesh screen. Larger drills passing through the one inch screen and held on the 1/2 x 3 inch screen may be counted without too much difficulty; the smaller drills passing through the finer screen are much more difficult to find. At first the material passing through the finer screen was examined in wet 1/2 bushel quantities, but this method was time consuming

100

and inaccurate. Later the material was air dried and shaken through nested screens, which facilitated drill counting and, though still laborious, proved more accurate. On October 16, 1953, a 20 minute run, representative of these observations, was made with the suction dredge over a piece of hard bottom; this was the fourth day of dredging on this particular piece of ground. During the 20 minute run 6 bushels of material retained on the 1/4 x 3 inch screen contained a total of 1,050 drills, and 45 bushels of material passing through this screen contained a total of 14,400 small drills'. The conspicuously higher proportion of small drills is evident and anticipated. Although the conventional oyster dredge does not agitate its load to the degree performed by shaker screens, the implication that small drills may be washed through the screen of the standard oyster dredge is further strengthened, particularly when the bottom is sandy and smaller materials readily wash out of the dredge or when the bottom is muddy and the dredge is washed vigorously to remove excess mud before being hauled aboard.

Cole (1942) on English oyster grounds confirms the fact that much of the damage done by drills passes unnoticed since it is principally spat that are drilled, and a drilled spat is very rarely seen because the drilled valve generally becomes detached and broken up soon after the spat gapes open. Thus the principal evidence of the activity of Urosalpinx is the failure of the spatfall to show up the following season. He notes similarly that it is practically impossible to detect the damage performed by freshly hatched drills among recently set spat, and this damage may be very great.

EUPLEURA CAUDATA

Because U. cinerea occurs more abundantly and is better known than Eupleura caudata, the bulk of this review is concerned with Urosalpinx. However, the more important aspects of the little known biology of Eupleura may properly be stressed here.

Galtsoff et al. (1937) report that adult Eupleura average 19 to 45 mm. in height.

There is considerable interest in the possibility that Eupleura may be increasing in abundance. Though the trend is not always consistent and may simply reflect a cyclical phenomenon. T. C. Nelson (1922) stated that in Little Egg Harbor, New Jersey, Eupleura constituted about 9.5% of the total drill population. In a count of 10,000 drills in Delaware Bay. J. R. Nelson (1931) observed that Eupleura comprised 2% of the total population. Sizer (1936) in the same bay found that out of a total of 16,200 drills collected, 7.4% were Eupleura, but that the distribution of both species varied greatly in different parts of the bay. Later he captured a total

101

of 130,070 drills, of which 1.6% were Eupleura. Stauber (1943). also in Delaware Bay. during 6 years of collecting noted the remarkable range of 0.03 to 35.50% Eupleura from station to station. In all cases but one his figures are based on annual catches of both species in excess of 10,000 and as high as 380,000. Over a period of years in which more than two million drills were taken, only 3.5% were Eupleura. Thus it is difficult to conclude, as did Haskin (1935) that the overall ratio of Eupleura to Urosalpinx is rapidly increasing on New Jersey oyster beds. It is possible nonetheless that in certain favorable areas the proportion of Eupleura is increasing. When Loosanoff (pers. com.), for example, started his researches in Connecticut in 1932 Eupleura was a rarity there; now in some areas in Long Island Sound he finds it as abundant, or more abundant, than Urosalpinx. And Andrews (pers. com.) in weekly collections of drills captured with baited drill traps on Wormleys Rock. York River, Virginia, finds in a total of 6,736 Urosalpinx and Eupleura collected during three summers in the last 12 years that the percentage of Eupleura increased as follows: 1942, 29.3%; 1948, 23.5%; and 1952, 82.0%. In Bogue Sound, North Carolina, in the vicinity of the Institute of Fisheries Research pier, Chestnut (pers. com.) in a general collection of 151 drills found only one Eupleura; and a collection in New River, North Carolina, of 23 drills contained two Eupleura.

T. C. Nelson (1922) in a study of the comparative destructiveness of Urosalpinx and Eupleura showed that when confined individually in cages with oysters, Eupleura destroyed slightly more oysters than did Urosalpinx. Haskin (1935) and Galtsoff et al. (1937) state that the two species of drills are about equally destructive; and the latter point out that Urosalpinx because of its much greater abundance is the more serious pest of the two species.

Haskin (1935) demonstrated that Eupleura is more active than Urosalpinx in egg deposition, laying an average of approximately 22 eggs per case, and a range of 8 to 42 eggs per case (based on a collection of 445 egg cases) in Barnegat Bay, New Jersey. Stauber (1943) noticed parallel fluctuations in the total number of egg cases of both species collected on drill traps throughout the season. High numbers of Eupleura egg cases were observed during the summer of 1937, and relatively lower concentrations were seen during the subsequent three summers.

Haskin (1935) first described a hinged cap on the egg case of Eupleura analagous to that on the Urosalpinx egg case through which the young drills escape.

Stauber (1943) without experimental evidence suggests that the two drills may possess similar minimum salinity death times

From preliminary field observations Loosanoff (pers. com.) is led to believe that Eupleura is better adapted to life on soft muddy bottom than Urosalpinx.

102

In one area near New Haven Harbor, Long Island Sound, where the bottom is extremely muddy and devoid of oysters, he finds in the summer that occasional shells of live clams (Mercenaria mercenaria) that protrude above the mud are completely covered with Eupleura egg cases and those of Urosalpinx are absent.

CONTROL
Introduction

A synthesis of the available knowledge on the biology of U. cinerea is a fundamental and fruitful prerequisite to an appraisal of methods concurrently employed in its control. The factors which contribute to the multiplication and distribution of the species (Federighi, 1931c) are strikingly manifested in such a review: (1) the presence of a hard calcareous shell and tightly fitting chitinous operculum which partially protect the drill from predators, unfavorable salinities and desiccation; (2) a singular freedom from enemies, except possibly its own kind; (3) a spawning behavior which provides protection and a food supply for the unhatched young in egg capsules; (4) young drills emerge from the egg case already covered by a strong shell and operculum and well equipped to plunder neighboring food organisms; (5) the capacity of a small proportion of young drills to "hitch hike" to adjacent ground on floating materials; (6) a high adaptability of the drill to a wide range of environ mental conditions; (7) the hibernation of a large proportion of drills buried in the sediment; (8) firm attachment of egg capsules on or in the vicinity of food organisms, especially oysters; (9) the fixed adherence of all post hatching stages to living oysters and shell and other substrates during summer and winter months; and (10) the relatively small size and drab coloration of the shell, particularly of recently hatched drills. The majority of these characteristics increase the chances for the dissemination of the species in the management of oysters.

Although the oyster drill appears to possess but few weaknesses in its life cycle, and thus far has rather successfully resisted man's relatively uncoordinated and sporadic attempts to control it, it does display certain characteristics which singly or collectively are or may be further employed with profit in combating it: (1) a limited low rate of migration for drill populations as a whole; (2) the tendency of drills to avoid soft muddy bottoms and sandy bottoms devoid of shell and other hard objects; (3) during the colder months of the year a relatively loose adherence of the foot of the drill to the substratum; (4) the possession by the drill of a positive geo-tactic response causing it to climb onto objects off the bottom at higher temperatures; (5) oviposition only at higher temperatures, on oyster beds on oysters and shell, on intertidal bottoms on rocks exposed at low tide; (6) the highly permeable nature of the egg case membranes to foreign ions; (7) a relatively long incubation period; (8) fatal effect of low salinities which are endured by the oyster; (9) a strong selective response of drills to the ectocrines of some prey, and possibly to the extracts of these; (10) larger size of the females than of males

103

In reviewing the possible methods of artificial control of the drill, T. C. Nelson (1922) lists poisoning, trapping, dredging, transplanting, and breeding of resistant oysters. Of these he considers poisoning impractical, and trapping not yet successful; but urges the development of poisoned baits. Since 1922 some progress has been accomplished in all of Nelson's suggestions except breeding of resistant oysters. In a recent personal communication he corroborates the writer's conviction that the evidence for drill dissemination by man is outstanding and conclusive and he suggests accordingly that the first procedure in oyster culture to undergo thorough overhauling is transplantation of oysters.

Federighi (1930b, 1931c) and Galtsoff et al. (1937) resolve the control of the drill into two distinct problems: the prevention of the migration and dissemination of the drill over new areas, and its removal and destruction from infested bottoms. Drill control in oyster cultural practices according to Galtsoff et al. falls roughly into mechanical and biological methods; the latter utilize knowledge of the life history and habits of the animal. Federighi emphasizes the fact that preventing the distribution of the drill to new areas, is somewhat easier than eradicating it once it has become established.

Stauber (1943) believes that drill control must be a continuous process by which Urosalpinx can be kept constantly in check. Thus large costs relative to a full scale program of drill control, should drills become numerous, may be prevented. He also states that no one aspect of drill control should be stressed at the expense of another, and that drill control should be carried out on grounds regardless of the lack of control of drills on adjacent bottom since mass migrations from ground to ground do not best explain the existing data on the distribution of drill populations. He also italicizes the important statement, which the writer would like to further emphasize, that in order to adequately determine the need for the application of control measures proper quantitative sampling of drills on grounds is mandatory.

Summarized in a general way, Stauber's recommendations for drill control in Delaware Bay may be abridged as follows: (1) before seed oysters are planted on a ground, resident drills should be removed; (2) if high drill concentrations appear after seed oysters are planted, oyster drill trapping should be employed from early April (to take advantage of the preoviposition period) to late October; (3) in replanting seed and adult oysters efforts should be made to remove accompanying drills; (4) if such transplantation is performed in May-July when oyster drill oviposition is at its height egg cases should be destroyed possibly by dipping or spraying with a selective toxic chemical; (5) when oysters are being dredged for market all accompanying drills should be segregated to reduce the need for subsequent drill removal from these grounds.

104

Cole (1951) believes that intensive dredging of heavily infested oyster ground during the period between spawning and hatching of drills will always be an important measure of control. At this time a very high proportion of females is collected by the dredge (Cole, 1941), possibly because the male is more easily detached from the cultch than the female. In addition egg cases which are usually deposited on objects large enough to be retained by the dredge are removed. T. C. Nelson (pers. com.) logically rebuts that the intensive dredging advocated by Cole would remove much of the new growth on the oysters, thus prolonging by one or more years the time required to market the crop, and subjecting the oysters to a longer period of attack by drills.

Production of oysters in the United States according to Glancy (1953) will never increase substantially until effective methods of drill control are universally established.

At this time no single panacea for the control of the drill is available. It is certain, though, that control must necessarily involve a number of cooperative, continuous, long range operations over a large area. The nature of the control may well have to vary from region to region commensurate with the depth of the water, the nature of the bottom, and the resources of those charged with the direction of the control operations. A critical review of all known methods of control which have been attempted is given in the following sections.

Capture of Drills and Egg Cases
Hand picking

The gathering of drills by hand in the intertidal zone at low water has been recommended by Cole (1942), Adams (1947), and Cahn (1950) as an effective means of control. Cole writes that on some beaches along the River Blackwater system in England one man can collect 500 drills in a few hours, and concludes that the detachment of men from normal dredging operations for the work of hand picking is justified. Adams advises that best results are obtained by picking early in the spring before egg cases are deposited. Cahn describes a novel program for drill eradication by hand picking in Japan where, to remove Japanese drills from intertidal oyster collectors, the prefectures of Hiroshima and Miyagi established an oyster drill extermination day at which time the school children swarm over the oyster beds at low tide and gather all the drills they can find.

T. C. Nelson (pers. com.) considers that hand picking of drills is theoretically sound, but is not as effective in actual practice as would be desired. On a recent trip to England, in company with Mistakidis and associates, he attempted hand picking drills on the intertidal flats of the River Roach. Even with utmost

105

care many oysters were pushed into the mud by their feet, raising the question whether such damage to oysters might not have equalled that which the collected drills would have inflicted. Nelson observes that on hard bottoms hand picking is more feasible, but since drills on exposed flats crawl under objects, careful hand picking would involve turning over every oyster and shell which lies sufficiently free of the bottom to permit drills to lodge underneath. He concludes that the drill trap offers the best known means for drill control in intertidal areas: it provides attractive food, cover from the sun, more favorable location for oviposition, and is easily tended.

So far as the writer knows the quantitative effectiveness of eradicating a small proportion of a drill population by such localized measures as hand picking, and by a number of other methods to be described, has never been tested. Such control measures probably bring only short term localized relief from predation. Although not verified experimentally in the case of drill populations, knowledge of other motile animal populations suggests that the extermination of a peripheral portion of a large population merely reduces the overall density and consequently inter- and intraspecies competition, with the result that the residual population soon resumes its former magnitude. Controlled tests should be effected before localized eradication measures on a costly and large scale basis are inaugurated.

Forks

The use of forks to shake drills from oysters on the deck of oyster boats is suggested by J. R. Nelson (1931), Federighi (1931c), and Galtsoff et al (1937) as a simple and inexpensive method. They also recommend that oysters be thrown overboard with forks instead of shovels so that drills present may pass through the tines onto the deck to be destroyed later. J. R. Nelson adds, however, that the number of drills so recovered is not large.

Concrete pillars

The use of small concrete pillars onto which drills will crawl has been suggested by Federighi (1931c) as a means of trapping. J. R. Nelson (1931) observes that these pillars will not collect drills as effectively as the drill trap, they expose a limited surface area, are heavy and difficult to handle in moderately deep water and in strong tidal currents. No field tests have been reported on this method; on the whole it appears quite impractical.

Oyster dredges

The conventional oyster dredge (J. R. Nelson, 1927) with the large mesh bag is quite inefficient for drill control. The greatest advantage is probably secured by the removal of shells and oysters bearing drill egg cases (Glancy, 1953). Cole

106

(1942; 1951) in trials made in English waters during dredging of oysters on muddy ground demonstrated that washing of the dredge to the extent necessary to free it from mud resulted in the loss of 75 to 100% of the drills. Mistakidis (1951), also in England, reports a similar experience. Cole points out that the inefficiency of the oyster dredge in collecting drills under three years of age is not too serious a matter as at first appears, for it is probable that the quantity of egg cases oviposited by small drills is insignificant in relation to that deposited by larger drills. In view of the high rate of destruction of young oyster spat by immature drills, Cole's thesis does not appear practical. Engle (1940) confirms reports that the use of a dredge chain bag with too large a mesh permits drills to escape, and adds that on the basis of the number of egg cases dredged in this way from year to year on a given bottom, the drills do not seem to be seriously reduced in number. According to Stauber and Lehmuth (1937) at least 10% of the drills collected in dredges with 1/2 inch mesh bags are washed out before the dredge load is dumped on deck, all the small and some of the medium sized drills being lost in this way.

The oyster dredge with a fine mesh bag was employed as early as the 1880's in Norwalk, Connecticut, to remove drills from hard bottoms, but failed to catch all the drills (Goode, 1884). Federighi (1931c) advocated the use of a small mesh bag on the oyster dredge for capturing drills, but stated that this method is effective only after all oysters have been harvested from drill infested oyster bottom, so that all the shell, drills, and debris so removed may be disposed of. Stauber (1943) notes that the 1/2 inch mesh bag holds mud easily and thus is not applicable on muddy bottoms, but it may be used successfully on harder bottoms in the manner of the Long Island Sound oystermen to remove the trash, including drills, after most of the oysters have been dredged. He concludes that this type of dredge is not only less efficient than the drill dredge but more costly to operate.

Deck screens

The method of drill control involving the screening of oysters on the deck of oyster boats during transplantation and harvesting is more efficient than forking (Federighi, 1931c; Galtsoff et al., 1937), and is the least expensive of the drill control methods but is less efficient than trapping (Stauber & Lehmuth, 1937). Drill infested oysters are shovelled against a one inch mesh screen of double weight chicken wire, or against an inclined or onto a horizontal perforated iron plate (Stauber & Lehmuth, 1937), and drills fall through the perforations. Exposure to air for a time loosens the drills and they tend to fall more readily from the oysters. The inclined screen, as compared to the horizontal screen, requires more handling with forks or shovels and thus is less desirable.

107

Stauber (1943) describes the use in Delaware Bay of a similar though stouter perforated metal plate placed horizontally under the dredge roller about four inches above the deck where it receives dredged oysters; in the process of culling or forking many drills fall through the screen and may be destroyed later. He concludes that the use of the deck screen in this way has proved even more spectacular in the transplantation of oysters, and in view of these results finds it difficult to understand why this control measure has not been applied more frequently; and that if methods of oyster culture in Delaware Bay are to continue along the pattern of the past few decades the deck screen is a valuable adjunct to drill control.

J. R. Nelson (1931; pers. com.) warns that hand shovelling of oysters onto an inclined screen as a result of which oysters tumble over it with considerable breakage is not to be recommended because such rough handling breaks off the thin bills of many of the oysters, resulting in the death of some and retardation of the growth rate of others. He adds that the horizontal deck screen as used in Delaware Bay does not injure oysters appreciably.

Drill dredge

This method of capture utilizes a wedge shaped dredge fitted at the top with an inclined screen which when dragged over drill infested oyster beds forces oysters over the dredge and drills automatically fall into the dredge pan. This type of dredge was invented by Captain T. Thomas in the late 19th century in the New Haven, Connecticut, area, where it was considered the most promising means of control at the time (Moore, 1898a). During the present century modified types have been employed in at least Chesapeake Bay and Delaware Bay. Federighi (1931c) and Galtsoff et al. (1937) give detailed drawings and descriptions of the device. Federighi concludes that the dredge is quite satisfactory provided the oyster population is not too dense, and states that the most effective time for dredging is in the early spring when the drills have moved onto the upper layers of oysters and shell and before spawning begins.

J. R. Nelson (1931) tested three different designs of drill dredges which he built, but results were not sufficiently encouraging to warrant the recommendation of the application of this method in Delaware Bay. He found that the dredge nicks the bills of oysters, soon chokes with bottom material, and does not capture enough drills to justify its use; but on beds from which oysters have been harvested and which are being prepared for a subsequent planting or for the catch of a set, as confirmed by Stauber and Lehmuth (1937), the dredge has some value

Galtsoff et al. (1937) corroborate J. R. Nelson's observations on the inefficiency of the dredge. A test in Delaware Bay on a 15 acre oyster bottom

108

employing a single dredge demonstrated a gradual though irregular decrease in the number of drills caught from 3,355 during the first hour to only 1,858 during the ninth hour. Stauber and Lehmuth (1937) suggest that it should be possible by two days of work on a 10 acre bottom to reduce the hourly catch of drills by 80%. Stauber (1943) later reported in more detail on controlled testing of the drill dredge in Delaware Bay. During tests performed on three separate grounds in April he found that the 1/2 inch mesh screen bag of an oyster dredge proved 14.5 times as efficient, and the drill dredge 50 times as efficient as the standard wide mesh bag of the oyster dredge in catching drills. He agrees with Galtsoff et al. that prolonged work on a ground with the drill dredge results in a reduction of drill catches, but never in the complete removal of all drills. Stauber recommends the use of the drill dredge principally during the colder months of the year when it is more efficient because drills then are less firmly attached to substrata. On the other hand at this time of the year a large proportion of the drills may be buried in the bottom and those buried in small depressions would be missed by this dredge The optimum time for dredging should be in the spring as suggested by Federighi.

Drill box traps

Brief mention should be made of a drill box trap which has been described by Galtsoff et al. (1937). It consists of a galvanized iron box into which drills enter at the ends through special openings covered with hinged metal strips which swing inward only and prevent the escape of the drills. Seed oyster bait is placed in the center of the box in a small screened container. These authors indicate that the chief advantage of this kind of trap is that it may be left on oyster beds for a month or more untended. No mention is made of its efficiency, or how soon fouling organisms and rusting reduce the functioning of the hinged doors.

Federighi (1931c) made an effort to test the efficiency of a variety of mollusk and fish meats as bait in attracting drills to traps in the field. Owing to scavengers and putrefaction the bait did not last long enough to allow the drills to react to it, even though a variety of cage types was tested.

Drill trapping

The drill trap consists of a bag of stout galvanized chicken wire approximately 12 x 15 inches in size partly filled with small bait oysters and shell which give added weight and furnish substrate to help hold drills when the trap is tended. These bags are attached to lines at intervals of about 8 feet on drill infested ground and are tended at 5 to 7 day intervals at which time the traps are lifted, shaken to dislodge the drills, examined for egg case clusters, and reset on the bottom. Drills, reflecting a negative geotaxis, particularly during the spawning season, and an attraction to the bait, particularly if it is collected in poor growing regions, climb into the traps.

109

Drill traps were first employed successfully for several years in drill control in Chincoteague Bay, Virginia (Stauber, 1938), and were later modified and extensively utilized in Delaware Bay, New Jersey, by J. R. Nelson (1931) and by Stauber (1943). J. R. Nelson used young oysters from the Cape Shore of Delaware Bay for bait. He noticed that on beds of seed oysters traps baited with the seed from the same bed gave moderately good results, but drills would not abandon seed oysters on the bottom for the same size oysters in the traps as readily as they deserted larger oysters or vacant bottom for seed oysters. He concludes that the drill trap is a practical, inexpensive, effective method of combating the drill.

Galtsoff et al. (1937) in a continuation of the drill trap work in Delaware Bay and its extension to other regions, undertook large scale experiments with a view to investigating the practicability and improving the construction of the traps testing the various baits, and learning the most efficient method of setting the lines of traps. They describe and illustrate the method of the construction of the trap and state that one to two year old oysters in clusters with mussels and barnacles, when these are present on oysters, make the most serviceable bait. On grounds where drills are not abundant and when a small number of traps are to be scattered over a large area they recommend fastening traps individually or in pairs to stakes or from buoys. Where infestations are high they recommend several methods of trap arrangement: (a) a series of zigzag lines over the area to be trapped, which they state covers a maximum territory and tends to attract drills from all directions; (b) placing a single line of traps at one end of a ground and periodically moving it until the area has been trapped; this is more economical but less efficient; (c) surrounding the ground with traps and moving a long line of traps about within the area; the former prevents reinfestation of a cleaned section; (d) the most successful procedure, especially at Chincoteague, Virginia, involved the anchoring of one end of a long line of traps in the center of a ground and shifting the position of the line about 15 degrees at intervals. They observed that where food is so abundant that drills cannot be lured by bait, as along barnacle covered rocky shores of New England, trapping by this method is impossible, but that it is highly efficient on drill infested ground from which oysters have been harvested and consequently on which the drill food supply has decreased.

The most prolonged and intensive study of drill control by the method of drill trapping has been performed by Stauber in Delaware Bay. Early in his work (Stauber & Lehmuth, 1937) he concluded that the drill trap is the most practical and efficient device for the removal of drills and egg cases from plantings of young oysters which it does not disturb, and presents the only attempt known at figuring the cost of drill control by means of drill traps. He and Lehmuth in cooperation with Captain Joseph Fowler of Bivalve, New Jersey, demonstrated the one outstanding controlled example available of the practical benefits of drill trapping (T. C. Nelson, 1939-40; pers. com; Stauber & Lehmuth, 1937). A 100 acre ground

110

planted with young oysters was divided into four equal parts, and one of the quarters was intensively drill trapped. When the oysters were harvested the quarter which had been trapped yielded appreciably more oysters than the other three quarters combined'. In the final report on his investigations, Stauber (1943) recommends the zigzag arrangement of trap lines, with the trap lines placed parallel to the flow of the tidal current and left in place for a season. Placing the trap lines across the current did not increase drill catches and greatly increased operating difficulties. He recommends the use of unculled rank inexpensive brack ish water or creek oysters, which are usually well covered with attached barnacles and ribbed mussels, as drill trap bait. In utilizing bait from poorer growing habitats Stauber was the first to employ bait which Haskin's (1940, 1950) studies demonstrated is most effective in attracting drills in the more favorable oyster growing areas. Stauber also recommends rebaiting the traps four times during a trapping season in Delaware Bay. This may extend over that part of the year in which water temperatures above 10°C and salinities above 15 o/oo occur (approximately April to October inclusively). Stauber concludes that total removal of oyster drills by means of the drill trapping method alone is not possible unless huge sums of money are expended, but that "what can be accomplished is that the density of these pests can be reduced to such a level that oystering will become consistently more profitable". One defect of the drill trap developed in Delaware Bay is its lack of a find screen bottom which would retain drills as the traps are raised through deep water and in rough weather (T. C. Nelson, pers. com.)

Newcombe et al. (1941-42) guided by the work of Stauber carried out some preliminary drill trapping studies in the York River area, Virginia, during the summer of 1942, but report no new information. At first Cole (1942) strongly recommended the use of drill traps in English waters since the spawning habits of English drills are the same as those in America; but later he (1951) reversed his opinion, stating that drill traps are not effective. In a personal communication he writes that his group is using other types of "Traps" successfully, but that the standard pattern described by Galtsoff et al. (1937) is not apparently suited to British conditions. He does not describe the types employed. Despite the fact that drill trapping appears feasible from an economic point of view (Stauber & Lehmuth, 1937) and that it can be superior to other methods of control in reducing young drill con centrations, there has been no widespread adoption of the method for the control of the oyster drill (Glancy, 1953). T. C. Nelson (pers. com.) adds the clarifying note that the failure of those oyster growers, who have attempted drill control by means of drill trapping, to attempt cost accounting, their aversion to extra work, and their tendency to gamble are chief factors in the failure of drill trapping to date.

Hydraulic suction dredges

By far the most promising device which has yet appeared for the control of all stages of the life cycle of the drill is the hydraulic suction dredge. This was first developed by the Flower brothers in the nineteen thirties (H. B. Flower, 1938, 1948; J. R. Nelson, 1948a, 1948b) principally to remove drills and other enemies from oyster beds. Since that time some 6 more suction dredges have been constructed (Glancy, 1953). These include a number of interesting modifications associated with the needs of the owners. Because of the versatility of these dredges and their usefulness in deeper water, they are utilized in a number of the operations involved in oyster farming. J. R. Nelson (pers. com.) informs me that the F. Mansfield & Sons Co. suction dredge, the "Quinnipiac" is employed principally in drill control, but because it is the most efficient means available it is also employed to a great degree in handling oysters and shells and other materials encountered in oyster farming.

Glancy (1953), whose concern is shared by the writer, seriously questions, however, whether in drill control the method of operation of the suction dredge has always been based upon the soundest biological principles. Among the many pertin ent questions which he poses are the following: (1) In the interests of maximum efficiency in drill control is it advisable to design suction dredges for a number of oyster cultural operations? (2) Are intake nozzles designed to most efficiently remove drills from grounds over which they operate? (3) Is the construction of the dredge boat adapted to effectively retain small drills? (4) What minimum width of strip should be cleaned around uninfested oyster grounds to prevent the migration of drills from adjacent infested areas? (5) What is the most expedient means of disposing of captured drills? (6) What is the effect of this kind of dredging on oyster bottom and on the animals living there? As is indicated in this review only some of these questions have received answers, and most of these are incomplete

It is worth noting that in actual practice, at least one company, the Frank M. Flower & Sons Co., has achieved satisfactory control of drills on oyster grounds by ingenious application of the company suction dredge, the "Frank M. Flower" (H. B. Flower, 1948). As developed by H. B. Flower and currently applied (H. B. Flower, pers. com.) this method of control involves three separate transplantations of drills; the following description and discussion of the method will be based on the drill cleaning of a 50 acre piece of bottom:

Step. 1. After marketable oysters are harvested, the ground is thoroughly suction dredged and the material so obtained is passed onto a conveyer belt screen with trapezoidal openings 1 3/8 x 1 3/4 inches. The finer components of this material (a total of 3, 750 cubic yards of sand, fine shell, drills, etc., in this instance) are flushed through the screen, transported to the nearest soft muddy

112

bottom unsuitable for oyster culture and there spread uniformly but leaving a drill free zone around the periphery. This procedure adds a layer of material over the mud bottom (1/2 foot thick over 4-5 acres in this case), which after the drills are removed in Step 2 makes this area suitable for oyster culture. Suction dredge samples taken around such drill dumping areas indicate that during the summer drills move less than 150 feet from the concentration even when surrounded by plantings of oysters from brackish water. While the suction dredge is in opera tion a constant check is maintained on the number of drills (those not washing through a 1/4 inch mesh screen) brought up by the dredge so that trends in drill density on the bottom may be followed closely. The screen is held under the stream of sediment and water flowing from the conveyer belt screen for two seconds, and the drills retained thereon are counted immediately. In ideal sampling it is im- portant to relate the number of drills collected to a specific unit of the bottom; this is not possible with this method, although it does give a rough indication of drill population trends. A further error in sampling is introduced by the variable speeds of the dredge over the bottom due to fluctuations in wind and tidal currents.

Step 2. A few weeks later the ground on which the drills were concentrated in Step 1 is given a thorough suction dredging and these sediments (375 cubic yards in this example) are transported to a third area of soft mud bottom where, this time, the drills and sand and shell are concentrated as much as possible in one large heap.

Step 3. In another few weeks the drill heap accumulated in Step 2 is care- fully suction dredged and the drills and sediments so obtained are concentrated on a small piece of bottom located along the shore in an intertidal area accessible to the dredge boat during high tide but separated from deeper water by mud bottom.

In the drill cleaning operation of the 50 acres described in the previous three paragraphs, Flower estimates that a total of five million drills were removed. This represents an approximate concentration of two to three drills per square foot of oyster bottom before cleaning. The scanty data available on the capture of drills by means of the suction dredge strongly suggests that the nozzle is not catch- ing all the drills present on the bottom during the first, or even the subsequent few dredgings. As F. B. Flower (pers. com.) suggests, the efficiency of the nozzle undoubtedly varies from boat to boat and from time to time on each boat depending on the type of bottom, current, weather conditions, and the attention given by captain and crew to the operation.

Besides effecting satisfactory control of the drill for the Frank M. Flower & Sons Co., this method, since its adoption, has contributed about 40 acres of new hard oyster bottom. It is also instructive to analyze the method from a biological point of view. The method takes advantage of (1) the partial barrier against migration which a soft mud bottom devoid of shell seems to impose on the drill; (2) the tendency of sand and fine shell fragments to sink into mud and simultaneously of drills to crawl to the surface, thus decreasing the volume of sediment to be transported at each step (for example, the volume of sediment moved during Step 1 was 3,750 cubic yards, and during Step 2, only 375 cubic yards); and (3) the destruction of drills buried deeply in the sediments of the layers of sediment tend to become reversed. It is not known from what depth of sediment drills may emerge and survive. These depths are undoubtedly related to sediment type and particle size, and size of drill. Small drills probably cannot emerge from as great a depth as large drills. The Flower method has the unusual merit of striking with equal, if not greater, severity at the younger drills. It is also evident in this method that egg cases, if dredging is performed during the spawning season, will have to be treated separately on oysters and shell not passing through the screen of the conveyer belt

In conclusion it may be suggested that for maximum returns in drill eradication procedures employing suction dredges, careful quantitative checks should be maintained, not only on the rate of removal of drills from grounds being dredged, but on the trends of drill densities on bottoms between periods of dredging. Since effective operation of the intake nozzle is fundamental to satisfactory operation of the dredge as a whole, periodic tests on its effectiveness are important. In addition, a more complete knowledge of the behavior and life history of the drill would unquestionably contribute much in the development of improved design and operation of these dredges. J. R. Nelson (pers. com.), for example, emphasizes the fact that information on the seasonal habits of the drill would aid materially in the development of improved nozzle design---considerable information of this kind is now available. It is evident that further research on both the dredges and the drill, on a cooperative basis, by dredge operators and marine biologists is to be recommended as an important step toward achieving maximum efficiency and economy in drill control.

Fate of drills passing through suction dredges

H. B. Flower (1948) has questioned the survival of drills which pass through a suction dredge. F. B. Flower, working at the Oyster Research Laboratory, New Jersey (pers. com.), has shown that such treatment does not seem to injure them. Forty-eight drills varying in height from 6.4 to 22 mm. and which passed through the 8 inch centrifugal suction pump of the "Luther Bateman" dredge in October, 1953, were placed in aquaria in the laboratory for observation. They were maintained in water of a salinity of 24 o/oo, at temperatures varying from 16.8 to 20.8°C, and 1953 oyster set was added for food. Twenty-five drills ranging in height from 15 to 25 mm. which had not passed through the dredge were set up as a control. During the three weeks of the experiment the large drills actively drilled and fed on the young oysters and the smaller drills were active but killed very few oysters presumably because the oysters were too large to be effectively drilled by them. In the entire experiment only one small drill (in the experimental aquaria) did not survive.

Destruction of Drills and Egg Cases

A wide variety of methods for the destruction of drills and their egg cases in situ on the bottom 'and after capture appear in scattered published and unpublished articles. These take the form of untried suggestions, or procedures tested with different degrees of completeness. Some of these methods are cited here only for their historical interest; others have merit in theory but are implausible economically and practicably, or both; and some methods, after research or after extended research and large scale testing, may result in wide application.

Desiccation

All stages of Urosalpinx within the egg case may be killed by exposure on land for three or four days (Galtsoff et al., 1937). The duration of exposure necessary to destroy adults is not known. It would be useful to know the lethal exposure time for drills of varying size in the shade and in the sun and at varying depths in bottom trash dredged with the drills over a wide range of atmospheric temperatures.

Heat
Hot water

Galtsoff et al. (1937) suggest that drills and egg cases collected during the most vigorous part of the spawning season on such devices as shells in wire bags, old tin cans, and on similar materials, be destroyed by dipping the collectors in boiling water for 10 to 15 seconds. This immediately kills the drills and young and permits immediate replacement of the collectors on the beds. There is great need for some means of destroying drills which pass through the vibrating screens of hydraulic suction dredges usually in company with large quantities of fine trash and sediments. It is possible that hot water or steam may some day be employed for this purpose. The problem is complicated, however, by the voluminous quantities of trash and sediment which accompany the drills and which quickly dissipate the heat necessary to effect destruction of the drills. To date the mechanical aspects of the problem remain unsolved. The potential effectiveness and inexpensiveness of the method should encourage eventual practical solution.

Flame

Large concentrations of drills frequently collect on intertidal rocky areas near drill infested bottom during the summer. Their destruction by flame from blow torches was started many years ago by Gibbs (pers. com.) in New England. With the substitution of a flame thrower for the blow torch, the method is practiced by at least one oyster grower on the east coast (J. & J. W. Elsworth Co.) and is reported to be in use in various infested intertidal regions on the west coast of the United States (Lindsay and McMillin, 1950). No report has been made on the effectiveness of the method in reducing drill densities on the subtidal oyster grounds nearest the burned intertidal areas. Such tests should be made before extensive control by flame is undertaken.

The usefulness of flame in effecting significant mortalities among drills which fall directly into the water has been questioned by J. R. Nelson (pers. com.). Recently R. C. Nelson (1953) was able to demonstrate that under summer conditions flaming produces mortalities of 94 to 97%. Using a total of 628 active adult drills and flame from a bunsen burner and later from a blow torch, he simulated controlled field operations in the laboratory. Flaming was performed on drills clinging to rocks immediately removed from sea water (salinity approximately 30 o/oo; temperature about 22°C) and on those exposed at atmospheric conditions for as long as 90 minutes. Flames varied in intensity from about half to full strength, and were directed on individual or groups of drills until they released and fell into sea water beneath, which occurred in 3 to 11 seconds. Drills flamed on algal covered rocks usually drew algae between the shell and operculum and were held in position, greatly increasing the mortality rate. Rate and extent of recovery of burned drills was checked in the laboratory in running sea water for one to two weeks.

Electricity

A promising field of research is presented by the response of drills to certain electrical currents. Lindsay and his associates (1953), pioneering in this field (Applegate et al., 1954), are attempting to determine whether Japanese drills (Tritonalia japonica) can be killed, guided, repelled, or otherwise controlled by electrical means. Their preliminary experiments were performed in 20 x 24 inch photographic trays in one inch of still sea water in the laboratory and primarily with older juvenile and adult drills (20.5 to 41.2 mm. in height). The water was taken directly from the laboratory sea water supply which has a normal salinity of 28.8 to 30.1 o/oo. Temperatures ranged from 12.5 to 14.0°C, and pH from 8.0 to 8.2 (Lindsay, pers. com.).

116

To date after preliminary trials they have found that at 14.3 milliamperes per square inch of direct current drills immediately withdraw into their shells and remain there while the current is on. In many cases they do not resume active crawling for several hours after the experiment has been terminated. At 5.9 to 14.3 milliamperes drills become very active. It is between these intensities that Lindsay's group seek a guiding or repelling effect. Attempts to kill drills with electricity have been unsuccessful. Preliminary experiments with electrical fences have also failed. No adverse effects on oysters have been noted.

Ultrasonics

Although no research has been performed on the utilization of ultrasonics in drill control (Henry, 1954), investigations in this direction may reveal a frequency, possibly of high intensity, to which Urosalpinx is specifically sensitive, and which may be exploited in repelling or attracting or even permanently inactivating the drill in situ on the bottom.

Fresh, brackish, and brine water

Federighi (1931c) and Galtsoff et al. (1937) recommend floating drill infested oysters in brackish water as an unusually efficient method for killing all stages of the drill, particularly during the transplantation of oysters. In practice drill infested dredged oysters are placed in large floats and anchored for about 10 days in brackish water which is fresh enough to kill the drills but salty enough to cause no damage to the oysters. Inasmuch as destruction of drills in brackish water depends on a specific low salinity, the salinity of the water in which the drills have lived, and the temperature of the water, these environmental factors should be determined for each region. It can be seen that this method will prove practic able only where brackish water is available in the near vicinity of drill infested oysters. J. R. Nelson (1931) states that the transplantation of drill infested oysters to grounds overlain by relatively fresh water provides a good method of drill control where it can be utilized. Lindsay and McMillin (1950) report that the technic of flooding diked oyster beds with fresh water to kill unhatched drills is in use in one place on the west coast of the United States and is economical and has apparently proved effective against further reproduction.

In Japan an interesting modification of these methods is suggested by Hori (quoted by Cahn, 1950) in control of Japanese drills. A combination of the following two methods, either of which will work singly but not as efficiently, is recommended: (1) Dissolve table salt at the rate of 6 grams per liter of sea water and immerse drill infested seed oysters in this solution for one to two minutes; after shaking off the drills, wash the seed in normal sea water; (2) immerse the seed oysters in fresh water for about two minutes and after shaking off the drills wash in normal sea water.

117

The degree of handling evident in this method does not lend itself to use in America.

Magnesium chloride

Hori (quoted by Cahn, 1950) to remove oyster drills from oyster spat also recommends placing the seed oysters in a 30% solution of magnesium chloride ($MgCl_2$) for one to two minutes; after shaking off the drills wash seed in normal sea water. He states that a combination of the brackish and fresh water treatment is more effective.

Copper sulfate

Engle (1941) reported the effective killing of the prehatching stages of the oyster drill in the laboratory in Connecticut by dipping the egg cases in a solution of one part of $CuSo_4$ in 200 parts of sea water for one minute. Development of the stages ceased shortly after exposure to the poison and all ages were destroyed. The oysters were unharmed by this treatment even after exposure of 10 minutes. Weaker solutions of the chemical, 1:300, did not kill all the embryos, but stronger solutions were as effective as 1:200. He recommended the use of a longer exposure time in large scale field control operations.

These experiments were repeated by Newcombe in Virginia (1941-42) and the results generally confirmed Engle's findings, but indicated that under the Virginia conditions dipping of commercial oysters bearing drill egg cases in a 1:500 solution of $CuSO_4$ for one minute was adequate.

Lindsay and McMillin (1950) report that $CuSO_4$ has been used to destroy drills on a commercial scale in Liberty Bay, west coast of the United States, on the recommendation of A. J. Bajkov. The chemical was applied as crystals mixed with a wetting agent and spread from the dusting hopper of an airplane. The reported results of this dusting indicate that the $CuSO_4$ was highly effective against egg case stages and young drills but only moderately effective against adults, and that no effects on oysters were noted. Lindsay and McMillin carried out a number of experiments in the laboratory in the State of Washington in 1945 and again in 1949 and also found that $CuSO_4$ has a decided toxic effect on the prehatching stages of the drill even in concentrations as low as 1:5000 to 1:400. They performed no tests on young hatched drills, but write that the efficiency of $CuSO_4$ in killing adult drills was not borne out by their tests. They also emphasize the fact that $CuSO_4$ is extremely toxic to young salmon and to minute plant and animal life, and although very little quantitative information is available on its effects on marine life, it is highly probable that $CuSO_4$ could seriously reduce the quantity of microscopic food in a bay where considerable quantities of this poison were used to kill oyster drills. The latter point is also stressed by T. C. Nelson (pers. com.). Lindsay and

118

McMillin conclude that $CuSO_4$ cannot be recommended for controlling oyster drills in the field where other commercial species have to be considered.

These studies indicate that $CuSO_4$ may have some application only as a dipping agent to destroy prehatching drill stages. The relatively small quantities of $CuSO_4$ which would be returned to estuarine waters in this method would very quickly be dissipated below toxic concentrations. On the other hand the high cost of handling may well negate its use in this way.

Mercuric chloride

According to Lindsay and his associates (Anon., 1948; Lindsay & McMillin, 1950; Lindsay, pers. com.) mercuric chloride ($HgCl_2$) is moderately effective in the control of the Japanese oyster drill. These investigators at first utilized the chemical in an attempt to eradicate drills among oysters in water in diked beds. Final concentrations ranging from one part of the poison in 10,000 to 100,000 parts of sea water proved effective in destroying drills and the young in egg cases in a period of three to four hours. However some mortalities occurred among the oysters, and tests made by the U.S. Food and Drug Administration of oyster meats from oysters exposed to the poison 7 months previously demonstrated the presence of small but abnormal amounts of mercury.

These results led to the application of $HgCl_2$ as a spray over intertidal grounds at low water. Eight pounds of $HgCl_2$ and 8 pounds of a wetting agent (such as Ultrawet) dissolved in 50 gallons of either fresh or sea water was sprayed over infested exposed oyster grounds on hot sunny days at a rate of 50 gallons per acre. This method proved quite effective in destroying unhatched drills, but killed only those adults submerged in small tidal pools. For a time several oyster companies used the method for destroying unhatched drills on the west coast of the United States. Because of a possible accumulation of mercuric chloride in oysters, its extreme danger to careless persons applying it, and its high cost, its use has been largely discontinued. Lindsay believes that on drill infested grounds which become exposed at low tide $HgCl_2$ properly applied, as to the outside of dike walls and to oyster grounds from which oysters have been harvested, could be quite effective in exterminating drills.

On the basis of these investigations, Korringa (1949) developed a similar method for the control of Crepidula which he says has since been adopted by Dutch oystermen with considerable success. Oyster collectors covered with Crepidula are dredged, brought ashore, washed thoroughly, placed in large concrete tanks, and immersed in sea water containing one part of $HgCl_2$ to 15,000 parts of water for two hours. Young oysters are said to close and to remain unharmed, but Crepidula and early stages of shell disease are destroyed.

119

Because of the toxic nature of mercuric chloride, research aimed at its use in eradication of a species should be performed with great caution and under strict quantitative control.

Formalin

Stauber (1943) looking for an organic compound which could be detoxified by organic matter after treatment, found that a solution of one part of formalin and 100 parts of sea water (approximately 1:300 formaldehyde, H_2CO) killed all egg case stages in three minutes. The same concentration for only one minute did not kill all the embryos and some of these formed atypically for a time, but none ever hatched. Field experiments in which a 1:100 solution of formalin was used for five minutes confirmed his laboratory data. In these studies oysters with attached egg cases were dipped and then confined in cages. Some oysters were killed by this treatment, particularly those whose bills were injured in any way. Stauber suggests that oysters with injured bills and exposed to the chemical are so weakened that they gape and then are vulnerable to the attacks of predators. He concludes that formalin is not as effective a molluscacide in this case as copper sulfate, and emphasizes the need for extensive field trials in the development of large scale control procedures.

Rotenone and amox

Newcombe and associates, Virginia Fisheries Laboratory (1941-42), performed a number of laboratory experiments to determine the efficiency of various concentrations of rotenone, $C_{23}H_{22}O_6$, in killing the various developmental stages of drills within the egg case. Egg cases attached to live oysters were submerged in rotenone solutions and then placed in flowing York River, Virginia, water of salinity 18 o/oo and temperatures averaging 25° C and watched daily until disintegration or hatching of the young was observed. Treatment was most effective if egg cases were permitted to remain out of water in the sun for at least two hours after treatment with rotenone, since increased temperature speeds the action of rotenone. It was determined that fresh solutions of rotenone in concentrations of 3:1,000 applied to the egg cases either as a spray or by dipping, effectively arrested development of all stages up to those in which the shell was beginning to form. No concentration of rotenone used was effective in killing later prehatching stages. A similar insecticide, amox, proved to have approximately the same effect as rotenone. Concentrations of rotenone and amox in all experiments proved nontoxic to oysters.

Search for new compounds

The possibility of controlling drills on the bottom where they occur through the application of chemicals has been in the minds of marine biologists for some time.

120

Haskin (1950) in work performed in 1935-1937 which demonstrated that chemical attraction plays a dominant role in food selection by drills, was able to suggest that his findings might provide clues in the discovery of effective baits for traps in the control of Urosalpinx.

Loosanoff, Nomejko, and Miller (1953) initiated the first and only known wide scale search for compounds which might be utilized in the control of U. cinerea. They began experiments in 1947 seeking substances which would prove inexpensive, affect oyster enemies injuriously, and remain harmless to man and to commercial marine organisms. To date over 1,000 compounds have been screened under meticulously controlled laboratory conditions. The screening procedure involves the incorporation of the test chemical in agar blocks which are then placed in trays of sea water with drills. The behavior of the drills is observed continuously during the experiment. Of the total number of compounds tested about 50 show promise as repellents and approximately 60 as attractors of Urosalpinx. After further testing the latter may be useful as bait in drill traps. These investigators also discovered several toxic compounds which in comparatively light concentrations cause the death of drills. A number of chemicals were discovered which may be useful in an indirect means of control. These upon addition in small quantities to sea water around drills cause the soft parts of the drills and of other gastropods to swell far out of the protecting shell. During this stage, which usually lasts several hours to several days, the snails are incapable of locomotion and of contraction and thus are ready prey to crabs which suffer no ill effects. Likewise oysters, clams, and mussels are unaffected by these relaxing substances. The valuable information accumulated by these researchers, though extensive, is based entirely upon preliminary laboratory experiments, thus no final recommendations are yet available. Numerous additional screening tests and actual field experiments are mandatory before large scale field applications are possible. The search for new compounds and the effect of these compounds on oyster drills is being continued (Loosanoff, pers. com.) at the U. S. Fish and Wildlife Biological Laboratory, Milford, Connecticut.

Physical and chemical barriers

J. B. Glancy, West Sayville, New York (pers. com.) has invented a device (for which he is making patent applications in a number of countries) which he describes as an "oyster seed collector and drill eradicator". The collector, by interposing a physical and a chemical barrier between drill infested bottom and elevated oyster seed, prevents the crawling of drills onto oyster seed.

Glancy's collector is constructed as follows: a wide spreading base of two cross bars of angle iron supports a central vertical pipe about a foot in height which in turn underpins a second set of shorter cross bars. About 12 chicken wire bags filled with cultch bars fastened on the elevated cross bare. A cylinder with one end open and directed downward lies at the junction of the upright support and the elevated cross bars. An extension of the upright pipe rises above and is guyed by wires to the ends of the elevated cross bars, and thus contributes further anchorage for the cultch bags. The top of the upright pipe is provided with a loop for receiving a line by which the collector is hoisted from and lowered into the water with a power winch.

The only means by which drills may reach the cultch on the collectors are over the supporting central pipe and the inner and outer surfaces of the cylinder. The latter is baffled internally to provide a labyrinthine passage, and since it is air tight and passes open end down into the water when the collector is lowered, it traps a pocket of air. The baffle reduces the exchange of water between the inside and the outside of the cylinder, and in combination with the air pocket probably effectively stops the crawling of drills onto the seed oysters above. Upon standing for a time the concentration of oxygen, particularly in the presence of decaying organic matter which would tend to accumulate or could be introduced at the start, is reduced in the air pocket, and such gases as methane, ammonia, carbon dioxide, and hydrogen sulfide would tend to accumulate. The surfaces of the cylinder and the lower portions of the collector are painted with antifouling compounds containing an algicide. These should further increase the effectives of the cylinder as a barrier to drill migration.

Glancy reports considerable success in obtaining oyster seed on his collectors in heavily drill infested areas during his first season of large scale trials in 1954, and is planning an expansion of his project. He finds that his collectors do keep drills away from oyster sets. His experiments were carried out for the most part in the waters around Robins Island in Peconic Bay, Long Island. Setting was light last summer in this area and the final counts gave 100 to 500 spat per bushel of cultch on the collectors while no spat survived on the surrounding bottom. Size of the spat, which is increased in the greater flow of water off the bottom, by the end of the summer averaged 1-1/4 inches with a maximum of 2 inches in length. He transplanted 2,000 bushels of this to other areas.

Clancy's collector should provide an unusually good means for determining to what degree young drills on adjacent bottoms will ride on floating debris to oyster seed in the collector. To date this has not occurred, or if it has, has taken place to so slight a degree that no noticeable effects have been observed. Clancy's method of drill control involves more handling and is more costly than the conventional method of broadcasting shell on the bottom. But because of the critical shortages of oyster seed in many areas it is becoming necessary to utilize methods such as Clancy's for the production of seed, and to place them on a paying practical basis.

Glancy also describes his collector as a possible drill eradicator. He suggests that a slightly soluble chemical which upon hydrolysis will release a toxic gas be placed in the pocket of air in the inner recesses of the baffled cylinder. He further suggests that a grease containing DDT, copper, mercury, and arsenic salts, or combinations of these, could be spread over the interior of the cylinder. Drills crawling over these would absorb lethal doses of the chemicals, or, irritated by them, would withdraw within their shells carrying lethal doses closed behind the operculum, and destruction would then be completed after the drills fell to the bottom. The baffles within the cylinder are said to effectively diminish the dissipation of these substances into the ambient water outside.

To what degree drills will be killed by moving into the cylinder even in the presence of high concentrations of poisonous salts and gases has not been investigated; nor is it reported whether dead drills have been found on the bottom under the cylinders. Urosalpinx possesses keen chemoreceptors as judged by their response to young oysters and it is likely that at the first contact with the peripheral areas of low densities of noxious substances they will retreat, and this initial dose may not prove inactivating or lethal. The response of drills to concentration gradients of these substances and their lethal effects should certainly be studied before the collector is used specifically as a drill eradicator. Great caution should be taken that whatever poisons are used in the cylinders do not pass into the sea water in quantities sufficient to harm other marine life or be incorporated in the tissues of oysters making them unfit for human consumption. This may occur, as has been pointed out, in the use of mercuric chloride, and possibly other poisonous metal salts. Toxic gases which soon dissipate in the water would probably be less harmful and more readily eliminated in the tidal circulation

It is probable that Glancy's seed collector will be most effective, not in destroying drills on the bottom, but in producing oyster seed free of drills through the first growing season. This in itself will be a major accomplishment and a long awaited contribution to oyster farming. In any event, studies on the possible use of the collector as an eradicator should be pursued.

Biotic

Cole (1951) writes that a possible biotic method of control of the oyster drill which his group is investigating is the multiplication of final hosts of certain trematode parasites which cause castration of the drill.

Chapman and Banner (1949) report that the drill may have a natural enemy in an unidentified amphipod which ordinarily lives with no apparent harm to the oysters in small tubes constructed on the outside of Olympia oyster shells. The amphipod enters the egg case of the drill through the operculum, possibly by burrowing. Within the egg case it constructs its mud tube. Whether it eats the young therein has not been determined, but no eggs or live drills were found in cases inhabited by the amphipod. In Oyster Bay, Washington, of 62 egg cases which were examined, 23 contained amphipods.

Utilization of Local Conditions
Low salinity

Since the activity of drills is suppressed by low salinities...the exact salinity value varying with the temperature of the water, duration of the low salinity, stage of the life cycle of the drill, and the previous salinity history of the drill...a fluctuating salinity barrier may be said to exist beyond which drills will not be found. According to Glancy (1953) this barrier affords a highly effective natural method of control which has been widely applied, particularly in the culture of oyster seed which thrive in the brackish drill free waters upbay from the salinity barrier. Applications of this method have been effective in such estuaries as Delaware Bay, New Jersey (Galtsoff et al., 1937; Stauber, 1943; Engle, 1953) and in the James River, Virginia (Engle, 1953). Drills tend to slowly repopulate these fringe areas whenever salinities above approximately 12-17 o/oo (seems to vary with the geographic location) and summer temperatures persist for a sufficient time, the former as a result of periods of low rainfall. T. C. Nelson (1922) and J. R. Nelson (1931) explain that in Delaware Bay it has been possible to build an industry yielding five million bushels of oysters annually only because of the existence of the natural normally drill free oyster seed beds upbay in water of low salinities. When the young oysters are transplanted to saltier water after about a year their shells have thickened sufficiently to afford more protection against the drill.

A discussion of the use of low salinities in the control of drills on oyster grounds is not complete without some mention of the effect of low salinities on oysters, since a difference in salinity of only a few parts per thousand may be sufficient to establish spawning beds of oysters protected from drills by the salinity barrier. Loosanoff (1952), in an important contribution on the behavior of oysters (Crassostrea virginica) in water of low salinities, provides this information. Using oysters dredged in waters of salinity 27 o/oo, maintained in running sea water of various concentrations and at various temperatures, he found that between 23 and 27°C only two oysters out of 50 died in a salinity of 7.5 o/oo in 30 days; at all salinities the rate of survival increased as the temperature decreased, and young oysters resisted unfavorable salinities as successfully as adult oysters. In a salinity of 7.5 o/oo oysters fed normally; started growing, though slowly; and normal gonad development took place, though oysters with ripe gonads spawned in a

124

salinity of only 5 o/oo. Loosanoff's lower salinity tolerance figure of 7.5 o/oo for oysters normally living in a salinity of 27 o/oo, when compared to the reported lower limit of 16 at summer temperatures for drills from Long Island Sound (Engle, 1953), suggests a relatively wide salinity range here tolerable to oysters but intolerable to drills. It is quite probable, as Loosanoff's data suggest, that oysters, like drills, normally living in lower salinities possess lower limits of salinity tolerance. These should be determined routinely for both drills and oysters so that the range of salinity, if present, within which oysters may be cultured free of drills may be known for specific oyster growing regions.

Mud and sand bottom devoid of hard objects

After a ground has been cleaned of drills and planted with seed oysters the new stock may be partially protected against the invasion of drills by the presence around the planting of a clean unplanted zone (Lindsay & McMillin, 1950; Cole, 1951; H. B. Flower, 1948). Andrews (pers. com.) notes that this may be effective simply because food organisms are far enough removed that drills do not detect them, rather than that drills are unable to cross such waste bottom. This is possible except where drills are located immediately up or down stream from food organisms. The addition of a continuous line of drill traps around the outer edge of the clean zone is recommended for further protection (Galtsoff et al., 1937). The use of a zone of bare ground in the control of drills by means of the hydraulic suction dredge by H. B. Flower (1948) has already been described in a previous section.

Temporary abandonment of bottom

Stauber (1943) has suggested that the procedure of permitting oyster grounds to "lie fallow" should be seriously considered as a part of commercial oyster management. He notes that bottom becomes relatively barren soon after oysters are dredged for market, and that the commoner organisms such as drills and boring sponge are greatly reduced in numbers; that colonization of a bottom with oysters appears to produce favorable conditions for the attachment and emigration of other animals, thus intensifying pest problems. By way of example he cites the case of a drill infested oyster bottom on which a moderate population of oysters had been raised from a natural spatfall. Without the application of drill control measures the owner next planted clean cultch, and obtained a good concentration of spat, but by fall all of these had been destroyed by drills. By contrast the same year a similar spatfall struck on an old vacant ground about 1-1/2 miles distant and by fall only 4% of the spat had been drilled. Stauber recommends that a bottom should remain unused for at least one year. T. C. Nelson (pers. com.), however, believes that the success of ground left "fallow" for a year or more is the exception rather than the rule. And Andrews (pers. com.) observes that if drills live four or five or more years, one year of fallowing is useless, particularly when a drill can live for almost a year without food.

Mistakidis (1951) confirmed Stauber's observation that Urosalpinx does not favor grounds in a poor state of cultivation. In his surveys he noted the close association of drills and grounds covered with considerable shell and in a fair state of cultivation. Cole (1951) adds that generally speaking derelict grounds overrun with Crepidula carry few Urosalpinx. On the other hand J. R. Nelson (pers. com.) finds on Fireplace, Long Island, that uncultivated bottoms overrun with Crepidula when cleaned with the suction dredge yield far higher drill counts than cultivated areas.

The fact that the drill is an omnivorous feeder and that dredging of oysters for market does not necessarily denude the bottom of all organisms or of all shells on which new organisms soon set or have set, casts further doubt on temporary abandonment of grounds as a means of drill control.

Removal of bottom trash

The practice of removal of old shells and debris from the bottom prior to oyster setting time, as done by some oyster farmers, and dumping it ashore where drills are killed by exposure is another effective means of combating the drill (Stauber, 1943; Engle, 1940). To counteract the objection that soft bottom in Delaware Bay does not permit this treatment, Stauber recommends the return of the original trash to the bottom after drying ashore. If the trash is not required on the bottom it can be eliminated by Flower's method of disposal. Control by trash removal is a sound one, and merits wider application. The method removes drills and egg cases of all ages, organisms which compete with oysters for the available food, other oyster pests such as Cliona and mud crabs, and the niches which harbor real and/or potential enemies of oysters.

Ratio of drills to prey

· Since in general drills approximate a rather uniform distribution over favorable bottom (Mistakidis, 1951), and since young oysters if planted sparsely over such bottom would tend to be more quickly destroyed than oysters planted thickly among the same concentration of drills, F. B. Flower (pers. com.; New Jersey Oyster Research Laboratory) has suggested as a temporary expedient that oyster seed be planted in maximum concentrations. In this manner the food requirements of the drills present are satisfied by only a partial destruction of the oysters, and overall short range survival of oysters is increased. At first glance this appears like an uneconomical means of control, but if no other is available, should prove useful as an expedient. Haskin (pers. com.) cites an application of this method of control in Delaware Bay. During the past four years such light natural sets of oysters have occurred in the upper less saline zone where Urosalpinx exist in Delaware Bay that Urosalpinx have destroyed them almost entirely. In October, 1953, 5,000 bushels of heavily spatted 1953 Cape Shore set (averaging about 2,000 spat per bushel) was transplanted to this area, and to date (November, 1954) the destruction of the spat by drills has been negligible.

Sharply pointed objects

In Japan Suehiro (1947) recommends the use of the chestnut burr on the rope of the collector string to prevent the climbing of the Japanese drill Rapana. He found that the soft footed drills would not cross the sharply pointed barriers.

Exposure on intertidal bottom

From preliminary studies in Seaside, Virginia, where he found that oyster setting occurs as high as three feet above low water mark but that destruction by drills is curtailed rather abruptly one to two feet above low water, Mackin (1946) suggests that in this region the drill may be controlled by utilizing the drill free zone for the culture of oysters. He admits that such high grounds are scarce, and that since the erection of artificial elevated surfaces is costly, available areas must be exploited to the fullest to compete economically with good natural subtidal grounds.

BENEFICIAL ASPECTS OF THE DRILL

Latham (1951) emphasizes in an exhaustive treatment of the ecology and economics of vertebrate predator management that the most destructive predators may be beneficial under certain circumstances. This seems to apply to U. cinerea. As Glancy (1953) points out, and with reason, the destruction of heavy sets of spat on marketable oysters is desirable since such sets render these oysters practically unmarketable. When dense sets do survive on adult oysters it is usually more profitable to handle the population as seed. He reports that in some of the southern states one of the drawbacks to oyster culture is the continual setting and survival of young oysters on older oysters, resulting in a product which, if used at all, can be handled only in canneries.

GENERAL DISCUSSION

One is led to believe from early reports (Ingersoll, 1881; Bur. Stat. N.J., 1902) that the oyster occurred in unusual abundance along the eastern coast of the United States during the 17th century. But mention of Urosalpinx is not found until

127

1822 (Say, 1822), and something of its destructiveness to oysters is not indicated until 1843 (DeKay, 1943). The writer suggests nonetheless, that drills have probably been predators of epifaunal bivalves like oysters since the evolution in Urosalpinx of the present drilling mechanism, and are as, or more, serious predators today than in early colonial times. The following four points support the latter hypothesis.

First, Urosalpinx, because of its small size, slow rate of movement, and inconspicuous method of predation is easily overlooked. This would explain why this snail went unnoticed by white man for a time and why it was not considered a serious oyster predator until recently.

Secondly, it is probable that modern oyster culture has tended to produce a hardier stock of drills. Oyster management practices over the decades have changed the aspect of oyster communities from the reef to the unistratal cultivated type, and in the course of this alteration have promoted a high degree of mixing of oysters and of their closely associated drill predators over wide areas. Such mingling should result in hybridization among interbreeding populations of Urosalpinx. Preliminary information on biological races indicates that such populations may cover extensive geographic areas, and that introduction and survival of new drills from distant estuaries is indeed a good possibility. It is also likely that mutations have occurred favoring further adaptation of the drill to the cultivated type of oyster community.

Third, the reported abundance of oysters in early colonial times does not necessarily indicate the existence then of less destructive drills. Stauber and T. C. Nelson (pers. com.) draw attention to a modern oyster reef in saline intertidal waters of Cape May, Delaware Bay, New Jersey, untouched commercially, which for many years has supported an unusually dense population of drills, yet has remained consistently productive. Stauber points out that each year new sets of oysters occur in sufficient abundance to protect older oysters and to permit survival of a portion of the new generation, and suggests that this biotic balance may have held in early colonial days. The inference here is that man in harvesting oysters automatically joins the depredatory forces of drills. He differs from drills, whose maximum damage is directed to the early stages of the oyster, in harvesting the older stages and in so managing his grounds that the setting of larval oysters on older oysters is reduced to a minimum. This procedure obviously reduces the buffering effect of young oysters, a dominant characteristic of productive oyster reefs, and encourages predation of older oysters by drills.

Fourth, the explanation given in the preceding paragraph explains the existence of oyster reefs in saline waters in the midst of drills; in zones of brackish water, the more characteristic environment of most oyster reefs in the early

128

colonial period. low salinity not only constituted the barrier to drills that it does today, but because of hydrographic conditions e x tent then, probably presented a more stable check over a greater area than at present. This may be elucidated as follows. In pre- and early colonial periods densely forested lands bounding the coastal waters served to feed a relatively constant supply of fresh water to estuaries, thus maintaining a wide relatively stable zone of brackish water the year around suited to oysters but intolerable to drills. With the advent of the white man and the consequent clearing of the land went much of the forest which had slowed the flow of water back to the sea. Now, principally in the spring, rain, melting snow and ice flood the estuaries at low temperatures at which reduced salinities inflict relatively little harm to drills. In the summer under conditions of reduced fresh water flow salinity in estuaries mounts, permitting drills to migrate farther upbay to previously uninfested oyster bottoms. With the invasion of saltier water upbay, oysters also tend to set farther upstream, but in general as estuarial shorelines converge toward fresh water the acreage of potential oyster producing drill free gound is also reduced.

It is questionable whether Urosalpinx exists in greater densities per unit area today than in precolonial times, but the expansion of oyster culture onto bottoms which did not previously support oysters, and the insidious invasion of drills into these new areas and other areas with oysters, strongly affirms the existence of a total greater number of Urosalpinx today than in early colonial times.

SUMMARY AND CONCLUSIONS

1. The earliest fossil shells of U. cinerea were collected in North Carolina and in Maryland in Miocene deposits approximately 28 million years old. The species is common along the Atlantic Coastal Plain of the United States in Pleistocene deposits approximately one million years old in a range extending from Florida to Massachusetts

2. Man in oyster cultural practices has unintentionally accelerated the mixing and dispersal of Urosalpinx so that today it is found broadly distributed along the eastern coast of North America from Canada to Florida, along the western coast of North America from Canada to California, and on the eastern coast of the British Isles. Its occurrence on the west coast of North America and in Great Britain represents introductions by man. Its centers of maximum density appear to extend along the east coast of the United States from Chesapeake Bay to Narragansett Bay. Bathymetrically it ranges from the mid intertidal zone to a depth of at least 120 feet.

3. An anatomical and functional description of the mantle cavity, nervous system, circulatory system, locomotory system, drilling and feeding organs,

excretory system, reproductive system, ova, and egg capsule helps to explain the success and high degree of adaptability of Urosalpinx to a wide range of ecological conditions.

4. Noticeable variation in the onset of spawning of the drill may reflect annual differences in spring water temperatures, physiological races, and/or incomplete information. The average number of egg cases oviposited per season varies from a few to 96 per drill, the number being larger in older drills. The average number of eggs per case varies from 8 to 12. Actual numbers range from 0 to 29, older drills ovipositing more than young sexually mature drills

5. Urosalpinx lacks a free swimming larval stage. Development occurs in the egg capsule. A prehatching mortality of 14 to 50% is reported. Duration of incubation is markedly influenced by temperature and may vary from 18 to 56 days. Newly hatched drills are fully shelled and capable of drilling small prey.

6. Growth rate data are incomplete. In America it is assumed drills reach a height of 8 to 19 mm. during the first growing season. In England drills are thought to attain a height of 10 to 20 mm. the first year, and to live to a maximum age of 13 to 14 years during which males may reach a height of 39 mm. and females 43 mm. Maximum heights of drills in America (longevity unknown) vary from 27 to 40 mm., and a giant subspecies reaches heights of 61 mm. Sexual maturity is said to be attained at ages varying from one to three years and at heights of 13 to 24 mm.

7. Urosalpinx displays some discrimination in its choice of food, but feeds upon a wide variety of animal prey, particularly young oysters, edible mussels, and barnacles. Anomia are only infrequently attacked. The drilling site is not limited to any specific region on the prey, nor necessarily to portions which are easier to penetrate. Although young oysters are attacked most commonly, those over 8 cm. in length are drilled by large Urosalpinx. It appears that the thickness of the shell is more important than the length in decreasing the rate of predation. The observation that the oyster drill secretes a toxic substance while drilling which kills its prey has not been confirmed. The number of prey destroyed by drills per given time increases as the size of the prey decreases, and larger drills destroy more prey than do smaller drills. In a temperature range of at least 13 to 24°C the rate of destruction increases with temperature; excessive exposure to air and to low salinities curtails drilling; drilling rate increases during the breeding season. The maximum average number of small oysters destroyed by small drills is recorded as 34 per week. On the average adult drills destroy oysters 4 to 6 cm. long at the rate of about 0.14 to 0.35 per week

8. Soft muddy bottoms, and to a lesser degree sandy bottoms, devoid of hard objects, are probably unfavorable for the growth, multiplication, and loco-motion of the drill.

9. At summer temperatures drill mortality rates increase rapidly as salinities fall, but this rate is markedly reduced as temperatures drop, so that at low winter temperatures drills can withstand unusually low salinities for pro-tracted periods. Minimum survival salinities at summer temperatures appear to vary from 12 to 17 o/oo in different regions.

10. Activities of Urosalpinx are noticeably influenced by temperature, and the initiation and cessation of these activities varies in different geographic regions. Locomotory movement takes place in different regions in a thermal range approx imately 2 to 10°C; feeding and drilling, 7 to 15°C; and oviposition 11 to 20°C.

11. At current velocities above 1.25 cm./sec. Urosalpinx turns into and moves against the current; at velocities below this no rheotaxis is displayed. A pronounced negative geotaxis is exhibited at temperatures approximately above 10°C. In strong light drills move away from the source of light; in dim light, toward it; and at weaker intensities the phototactic response is lost completely. Chemical attraction plays an important role in food selection by the drill.

12. A number of physiological races and at least two morphological sub-species of Urosalpinx cinerea occur.

13. The oyster drill is preyed upon by its own kind, by Polinices, Asterias, and possibly other animals, but the degree of predation is probably slight. It is host to at least three parasites

14. The bulk of the drill population probably migrates only to a limited degree, particularly over oyster bottoms; occasional exceptions may be explained on the basis of phoresis. The majority of drills on firm bottom devoid of oysters tend to move at an average rate of 15 to 24 feet per day in the direction of food. An unknown proportion of drills in populations near shores migrate intertidally to spawn. The "sudden appearance" of high concentrations of drills on oyster bottoms may be explained on the basis of incomplete removal of young drills and the subsequent growth of these, rather than on the basis of mass migration alone.

15. Eupleura caudata, close relative of U. cinerea, generally constitutes only a small percentage of the drill population within the range of these two species, but may be increasing in certain favorable areas. It is more active in oviposition than Urosalpinx, ovipositing an average of about 22 eggs per case.

16. Hand picking, forks, concrete pillars, oyster dredges, deck screens, drill dredges, drill box traps, drill trapping, and hydraulic suction dredges have been employed in the capture of oyster drills. Of these methods the hydraulic suction dredge seems the most promising.

17. A number of physical and chemical methods have been suggested and others employed in the destruction and exclusion of drills and their egg capsules. No one method is applicable in all circumstances. Exposure of the pest on shore and dumping on submerged barren muddy bottom appear to be the most inexpensive methods applied to date. The former method is entirely effective; the latter has not been adequately tested.

18. A number of local ecological conditions occur in various regions which have proved, or may prove, useful in drill control: low salinity, areas of barren mud, removal of bottom trash, and exposure on intertidal bottoms.

19. Although Urosalpinx is considered a menace principally to oyster culture, the presence of a limited few on bottoms supporting marketable oysters may be desirable in elimination of oyster set on these oysters.

20. Drills have probably been predators of exposed bivalves since the evolution in Urosalpinx of the present drilling mechanism, and are as, or more, serious predators today than in early colonial times. There is no evidence to indicate that the drill exists in greater densities per unit area today, but because of its widespread distribution exists in total greater numbers, than in precolonial times.

ACKNOWLEDGMENTS

I am greatly indebted to J. D. Andrews, A. F. Chestnut, J. B. Engle, H. H. Haskin, C. E. Lindsay, V. L. Loosanoff, I. W. Sizer, J. L. McHugh, P. S. Galtsoff, and L. A. Stauber for the use of unpublished data and reports without which this review would have been less complete. I am especially grateful to Dr. Thurlow C. Nelson who first stimulated and has continued to foster my interest in the oyster drill, and to Dr. Leslie A. Stauber for permission to fully abstract his important unpublished work on the biology and control of the drill in Delaware Bay, New Jersey, which represents many years of both laboratory and field investigations

The following investigators have taken time to correspond with me concerning various aspects of this review: R. T. Abbott, J. D. Andrews, P. A. Butler, A. F. Chestnut, W. J. Clench, H. A. Cole, J. H. Day, J. B. Engle, F. B. Flower, H. B. Flower, P. S. Galtsoff, H. N. Gibbs, J. B. Glancy, F. Haas, D. A. Hancock,

G. D. Hanna, H. H. Haskin, J. W. Hedgpeth, A. E. Hopkins, H. J. Humm,
C. E. Jenner, Mrs. van der Feen-van Benthem Jutting, P. Korringa, C. E.
Lindsay, V. L. Loosanoff, G. R. Lunz, I. D. Marriage, J. L. McHugh, M. N.
Mistakidis, J. R. Nelson, T. C. Nelson, H. G. Orcutt, H. A. Pilsbry, L. R.
Pomeroy, H. G. Richards, J. E. Rogers, L. A. Stauber, and D. E. Wallace.
My sincere thanks are extended to them.

My genuine appreciation is expressed to J. D. Andrews, A. F. Chestnut,
J. B. Engle, H. A. Cole, H. H. Haskin, V. L. Loosanoff, J. L. McHugh, J. R
Nelson, T. C. Nelson, P. S. Galtsoff, L. A. Stauber, and B. H. Willier for
generously checking the entire manuscript of this paper, in particular for errors
and misinterpretations of data. All commentary has been written by the author.

Library research for this paper was performed in the libraries of Rutgers
University, Princeton University, the American Museum of Natural History, and
the University of North Carolina at Chapel Hill. May I acknowledge with thanks
the genial cooperation of the staffs in these libraries. A preliminary draft of this
review was prepared while I was a member of the Department of Zoology, Rutgers
University, New Brunswick, New Jersey. Final writing and revision and publishing
was accomplished after I joined the staff of the Department of Zoology, University
of North Carolina, Chapel Hill, North Carolina. I acknowledge my appreciation
to both institutions for the many courtesies extended to me during the long prepara
tion of this monograph

What measure of accuracy and completeness this synthesis attains is in
large part the product of the cooperation which has been extended so generously by
so many. It is a pleasure to acknowledge this support.

LIST OF LITERATURE

Abbott, R. T.
    1954.   American Sea Shells. D. van Nostrand Co., Inc., N. Y. 541 pp.

Adams, J. R.
    1947.   The oyster drill in Canada. Fish. Res. Bd. Canada, Progr.
             Repts. 37: 14-18.

Allee, W. C., A. E. Emerson, O. Park, T. Park, K. P. Schmidt.
    1949.   Principles of Animal Ecology. W B. Saunders Co., Pa.

Anon.
    1948.   New poison controls oyster pests. Comm. Fish. Rev. 10: 35-36.

Applegate, V. C., P. T. Macy, V. E. Harris.
    1954.   Selected bibliography on the applications of electricity in fishery
             science. Spec. Sci. Rept.: Fish. No. 127: 1-55, U. S. Fish and
             Wildlife Service

Arey, L. B., and W. J. Crozier.
    1919.   The sensory responses of Chiton. Jr. Exp. Zool. 29: 157-260.

Baker, Bernadine B
    1951.   Interesting shells from the Delmarva Peninsula.
             Nautilus 64: 73-77.

Brooks, W. K.,
    1879-1880. Preliminary observations on the development of the marine
             prosobranchs. Studies Biol. Lab., Johns Hopkins Univ.,
             Chesapeake Zool. Lab. Sci. Results 1878: 121-142.

Buchanan-Wollaston, J. H., and W. C. Hodgson.
    1929.   A new method of treating frequency curves in fishery statistics,
             with some results. Jr. Cons. Int. Explor. Mer. 4: 207-225.

Bullock, T. H.
    1953.   Predator recognition and escape responses of some inter-
             tidal gastropods in the presence of starfish. Behavior 5: 1-11.

Bumpus, H. C.
1898.   The breeding of animals at Woods Hole during the month of
          May, 1898. Science, N. S. 8: 58-61.

Bureau of Statistics of New Jersey.
1902.   The oyster industry in New Jersey.
          Bur. Statistics, N. J., 25th Ann. Rept. 295-360.

Cahn, A. R.
1950   Oyster culture in Japan. Fish. Leaflet, Washington 383: 1-80.

Carpenter, H. F.
1902.   The shell-bearing Mollusca of Rhode Island. Nautilus 15:
          130-132.

Carriker, M. R.
1943.   On the structure and function of the proboscis in the common
          oyster drill, Urosalpinx cinerea Say. Jr. Morph. 73: 441-498.

1951.   Observations on the penetration of tightly closing bivalves by
          Busycon and other predators. Ecology 32: 73-83.

1953.   A review of those aspects of the biology of the oyster drill
          Urosalpinx cinerea (Say) fundamental to its control.
          Conv. Papers, Nat. Shellf. Assoc., New Orleans, La.: 51-60.

1954.   Seasonal vertical movements of oyster drills. Proc. Nat.
          Shellf. Assoc. 1954 (in press), 9 pp.

Chadwick, G. H.
1905.   Shells of Prince Edward Island. Nautilus 19: 103-104.

Chapman, W. M., A. H. Banner
1949.   Contributions to the life history of the Japanese oyster drill
          Tritonalia japonica, with notes on other enemies of the
          Olympia oyster, Ostrea lurida. Biol. Bull., No. 49-A,
          Sta. Wash., Dept. Fish. 169-200

Chestnut A. F., and W. E. Fahy.
1953.   Studies on the vertical distribution of setting of oysters in
          North Carolina. Proc. Gulf & Caribbean Fish. Inst., 5th
          Ann. Session, Nov., 1952: 106-112

135

Coe, W. R.

    1912.   Echinoderms of Connecticut. Bull. Conn. Geol. & Nat. Hist. Surv. 19: 1-152.

    1949.   Divergent methods of development in morphologically similar species of prosobranch gastropods. Jr. Morph. 82: 383-400.

Cole, H. A.

    1941.   Sex-ratio in Urosalpinx cinerea, the American oyster drill. Nature, Lond. 147: 116-117.

    1942.   The American whelk tingle, Urosalpinx cinerea (Say), on British oyster beds. Jr. Mar. Biol. Assoc. U. K. 25: 477-508.

    1951.   The British oyster industry and its problems. Rapp. Cons. Explor. Mer. 128: 7-17.

Collins, J. W.

    1891.   Notes on the oyster fishery of Connecticut. Bull. U. S. Fish. Comm. 9: 461-497.

    1892.   Report on the fisheries of the Pacific Coast of the United States. U. S. Comm. Fish & Fisheries, Rept. Comm. for 1888 (1892): 3-269

Dall, W. H.

    1889   A preliminary catalogue of the shell-bearing marine mollusks and brachiopods of the southeastern coast of the United States, with illustrations of many of the species. Bull. U.S. Nat. Mus. 37: 1-221.

    1907.   Notes. Nautilus 21:91.

    1921.   Summary of the marine shell-bearing mollusks of the northwest coast of America, from San Diego, California, to the Polar Sea. Bull. U. S. Nat. Mus. 112: 1-217.

Dean, B.

    1890.   The physical and biological characteristics of the natural oyster grounds of South Carolina. Bull. U. S. Fish. Comm. 10: 335-361.

DeKay, J. E
1843.   Mollusca and Crustacea of New York, Natural History of New
York: Zoology of New York, or the New York Fauna. Part V.
Mollusca. Carroll & Cook, Albany, N. Y.

Elsey, C. R.
1933.   Oysters in British Columbia. Bull. Biol. Bd. Canada 34: 1-34.

Engle, J. B.
1935-
1936.   Preliminary report of the U.S. Bureau of Fisheries oyster
drill control project in New Jersey, 1935-1936. Unpub.
Rept., U. S. Bur. Fish., Washington, D. C.

1940.   The oyster drills of Long Island Sound. Conv. Addr., Nat.
Shellf. Assoc., New Haven, Conn.

1941.   Further observations on the oyster drills of Long Island Sound,
with reference to the chemical control of embryos. Conv. Addr.,
Nat. Shellf. Assoc., Atlantic City, N. J.

1942.   Growth of the oyster drill, Urosalpinx cinerea Say, feeding
on four different food animals. Anat. Rec. 84: 505 (abstract).

1953.   Effect of Delaware River flow on oysters in the natural seed
beds of Delaware Bay. Rept. U. S. Fish & Wildl. Serv.,
Washington, D. C. 1-26 (limited distr.).

Federighi, H.
1929.   Rheotropism in Urosalpinx cinerea Say. Biol. Bull. Mar.
Biol. Lab. 56: 331-340.

1930a.   Salinity and size of Urosalpinx cinerea Say. Amer. Nat.
64: 183-188.

1930b.   Control of the common oyster drill. U. S. Bur. Fish., Econ.
Cir. No. 70: 1-7

1931a   Salinity death points of the oyster drill snail Urosalpinx
cinerea Say. Ecology 12: 346-353.

1931b   Further observations on the size of Urosalpinx cinerea Say.
Jr. Conchol., Lond. 19: 171-176.

137

Federighi, H.
    1931c.   Studies on the oyster drill (Urosalpinx cinerea Say).
             Bull. U. S. Bur. Fish. 47: 83-115.

Field, I. A.
    1921-
    1922.   Biology and economic value of the sea mussel Mytilus edulis.
             Bull. U. S. Bur. Fish. 38: 127-259.

Flower, F. B.
    1954.   A new enemy of the oyster drill. Science 120(3110):
             231-232.

Flower, H. B.
    1938.   Uses of the suction dredge and "stardust". Conv. Addr.,
             Nat. Shellf. Assoc. Providence, R. I.

    1948.   Mechanizing the cultivation of oysters: new Flower oyster
             dredge. Conv. Addr., Nat. Shellf. Assoc., Asbury Park, N. J.

Ford, J.
    1889.   List of shells of the New Jersey coast south of Brigantine
             Island. Nautilus 3: 27-29.

Fraser, J. H.
    1930-
    1931.   On the size of Urosalpinx cinerea (Say) with some observa-
             tions on weight-length relationships. Proc. Malac. Soc.,
             Lond. 19: 243-254.

Fretter, Vera.
    1941.   The genital ducts of some British stenoglossan prosobranchs.
             Jr. Mar. Biol. Assoc., U. K. 25: 173-211.

    1946.   The pedal sucker and anal gland of some British Stenoglossa.
             Proc. Malac. Soc., Lond. 27: 126-130.

Galtsoff, P. S., H. F. Prytherch, and J. B. Engle.
    1937.   Natural history and methods of controlling the common oyster
             drills (Urosalpinx cinerea Say and Eupleura caudata Say).
             U. S. Bur. Fish. Cir. No. 25: 1-24.

138

Galtsoff, P. S., and W. A. Chipman, Jr.,
1940. Oyster investigations in Maine. Unpub. Rept., U. S. Fish & Wildl. Serv., Washington, D. C.

Galtsoff, P.S., W. A. Chipman, Jr., J. B. Engle, and H. N Calderwood.
1947. Ecological and physiological studies of the effect of sulfate pulp mill wastes on oysters in the York River, Virginia. U. S. Fish & Wildl. Serv. Fish. Bull. 43: 59-186.

Galtsoff, P. S.
1954. Recent advances in the studies of the structure and formation of the shell of Crassostrea virginica. Proc. Nat. Shellf. Assoc. (in press).

Gardner, A. H.
1896. Dredging in Long Island Sound. Nautilus 9: 119-120.

Gardner, Julia.
1948. Mollusks from the Miocene and lower Pliocene of Virginia and North Carolina, Part 2 Scaphopoda and Gastropoda. U. S. Geol. Surv. Prof. Paper 199B: 178-310.

Glancy, J. B.
1953. Oyster production and oyster drill control. Conv. Papers, Nat. Shellf. Assoc., New Orleans, La.

Goode, G. B
1884. The Fisheries and the Fish Industries of the United States. Section I. Natural history of useful aquatic animals, with an atlas of 277 plates. U. S. Comm. Fish & Fish., Washington, D. C.

Gould, A. A.
1841. Report on the Invertebrata of Massachusetts, Comprising the Mollusca, Crustacea, Annelida, and Radiata. Folsom, Wells, and Thurston, Cambridge, Mass.

1870. Invertebrata of Massachusetts. 2nd. ed., Wright & Potter, Boston

Hackney, Anne G.
1944. List of Mollusca from around Beaufort, North Carolina, with notes on Tethys. Nautilus 58: 56-64.

Hanna, G. D.
   1939.   Exotic Mollusca in California. Bull. Dept. Agr., Sta.
            Calif. 28: 298-321.

Haskin, H. H.
   1935.   Investigations on the boring and reproduction activities
            of oyster drills Urosalpinx cinerea Say and Eupleura sp.
            Unpub. Rept. U. S. Bur. Fish., Washington, D. C.

   1937.   Studies on the movements of the common oyster drill,
            Urosalpinx cinerea (Say). Unpub. Rept., N. J. Oyster Res. Lab.

   1940.   The role of chemotropism in food selection by the oyster
            drill, Urosalpinx cinerea Say. Anat. Rec. 78: 95 (abstract).

   1950.   The selection of food by the common oyster drill, Urosalpinx
            cinerea Say. Proc. Nat. Shellf. Assoc. 1950: 62-68.

Hedgpeth, J. W
   1953.   An introduction to the zoogeography of the Northwestern
            Gulf of Mexico with reference to the invertebrate fauna.
            Publ. Inst. Mar. Sci. 3: 108-224.

Henderson, J. B., and P. Bartsch
   1915.   Littoral marine mollusks of Chincoteague Island, Va.
            Proc. U. S. Nat. Mus. 47: 411-421.

Henry, G. E.
   1954.   Ultrasonics. Sci. Amer. 190(5): 54-63.

Higgins, E.
   1940.   Progress in biological inquiries, 1939. Admin. Rept. 39,
            App. I, Rept. U. S. Comm. Fish. 1940: 1-96

Ingalls, R. A., and A. W. H. Needler.
   1942.   Survey of the shore mollusc resources of the Northumberland
            Strait Coast of Nova Scotia. Progr. Repts. Atlantic Coast
            Stas. No. 32, Atlantic Biol. Sta. Note No. 87: 8-10.

Ingersoll E.
   1881.   A report of the oyster industry of the United States. Sec.
            X, Monogr. B, Dept. Int., 10th Census U. S., Washington, D.C.

Jacot, A.
    1924.   Marine Mollusca of the Bridgeport, Connecticut, region.
               Nautilus 38: 49-51

Jenner, C. E.
    1954.   Photoperiodism in marine animals. Jr. Tenn. Acad. Sci.
               29(3): (abstract).

Jensen, A. S.
    1951.   Do the Naticidae (Gastropoda Prosobranchia) drill by
               chemical or by mechanical means? Vidensk. Medd. fra
               Dansk naturb. Foren. 113: 251-261.

Johnson, F.
    1942.   The Boylston Street fishweir. A study of the archaelogy,
               biology, and geology of a site on Boylston Street in the Back
               Bay district of Boston, Massachusetts. Papers Robert S.
               Peabody Found. Archaeol. 2: 1-212.

Johnson, C. W.
    1890.   An annotated list of the shells of St. Augustine, Fla.
               Nautilus 3: 103-105.

    1915.   Fauna of New England. 13. List of the Mollusca.
               Occ. Papers Boston Soc. Nat. Hist. 7: 1-231.

    1928.   Urosalpinx cinerea Say in England. Nautilus 42: 68.

Kellogg, J. L.
    1901.   Observations on the life-history of the common clam, Mya
               arenaria. Bull. U. S. Fish. Comm. 19: 193-202

Korringa, P.
    1949.   Nieuve aanwijzingen voor de bestrijding van slipper en
               schelpziekte. Visserijuieuws 2: 90-94.

    1951.   The shell of Ostrea edulis as a habitat. Arch. Neerlandaises
               Zool. 10: 31-152.

    1952.   Recent advances in oyster biology. Quart. Rev. Biol. 27:
               339-365.

Latham, R. M.
    1951.   The ecology and economics of predator management. Rept. II,
               Penna. Game Comm., Harrisburg (Final Rept. Pittman-
               Robertson Proj. 36-R): 1-96.

Lindsay, E. E., and D. C. McMillin.
    1950.   Control of Japanese oyster drills. Sta. Wash. Dept. Fish.,
               Puget Sd., Oyster Bull., Ser 9(1): 1-5.

Lindsay, C. E., R. E. Westley, and C. E. Woelke.
    1953.   Drill control. Sta. Wash. Dept. Fish., Puget Sd. Oyster Bull.,
               Ser 12(4): 12.

Loosanoff, V. L.
    1945.   Effects of sea water of reduced salinities upon starfish,
               A. forbesi, of Long Island Sound. Trans. Conn. Acad. Arts
               & Sci. 36: 813-835.

    1952.   Behavior of oysters in water of low salinities. Conv. Addr.
               Nat. Shellf. Assoc., Atlantic City, N. J.: 135-151.

    1953.   Fishery Biol. Lab., Milford, Conn., Bull. 17(1).

Loosanoff, V. L., and H. C. Davis.
    1950-
    1951.   Behavior of drills of different geographical districts at the same
               temperatures. Unpub. Rept. U. S. Fish & Wildl. Serv.,
               Milford, Conn.

Loosanoff, V. L., C. Nomejko, and W. Miller.
    1953.   Brief summary of experiments on screening chemical compounds
               for control of enemies of commercial mollusks. Unpub. Rept.
               U. S. Fish & Wildl. Serv., Milford, Conn.

Lucas, C. E.
    1947.   The ecological effects of external metabolites. Biol. Rev. 22:
               270-295.

Mackin, J. G.
    1946.   A study of oyster strike on the Seaside of Virginia. Contr.
               Va. Fish. Lab. No. 25, 18 pp.

Mazyck, W. G.
    1913.   Catalog of the Mollusca of South Carolina. Contr. Charleston
               Museum No. 2

McDermott, J.
    1952.   Ecological notes concerning the "Toadfish" Opsanus tau (L)
               in Delaware Bay during the summer of 1952. Unpub. Rept.,
               Oyster Res. Lab., N. J

McDougall, K. D.
    1943.   Sessile marine invertebrates of Beaufort, N. C. Ecol.
               Monogr. 13: 321-374.

Mead, A. D.
    1900.   The natural history of the star-fish. Bull. U. S. Fish Comm.
               1899: 203-224.

Metcalf, M. M.
    1930.   Salinity and size. Science, N. S. 72: 526-527.

Mistakidis, M. N.
    1951.   Quantitative studies of the bottom fauna of Essex oyster
               grounds. Min. Agr. & Fish. Investig. Ser. II, 17: 1-47.

Moore, H. F.
    1898a.   Oysters and methods of oyster culture. Rept. U. S. Comm.
               Fish & Fish. 1897.

    1898b.   Some factors in the oyster problem. Bull. U. S. Fish Comm.
               1897.

    1911.   Condition and extent of the natural oyster beds of Delaware.
               Bur. Fish. Doc. No. 745

    1912.   Enemy of the oyster, the drill, investigated by U. S.experts.
               Rept. Bd. Shellf. N. J. 1912: 71-75.

Needler, A. W H.
    1941.   Oyster farming in Eastern Canada. Bull. Fish. Res. Bd.
               Canada 60: 1-83.

Nelson, J
    1893.   Report of the biologist for 1892. Ann. Rept. N.J. Agr. Exp.
               Sta. for 1892: 205-271.

Nelson, J. R.
　1927.　Some principles of oyster dredging. Bull. N. J. Agr. Exp. Sta. 443: 1-21.

　1931.　Trapping the oyster drill. Bull. N. J Agr Exp. Sta. 523: 1-12.

　1948a.　Mechanizing the cultivation of oysters: recent developments and improvements in oyster dredges. Conv. Addr. Nat. Shellf. Assoc., Asbury Park, N. J.

Nelson, J. R., H. B. Flower, and A. E. Hopkins.
　1948b.　Mechanization of oyster cultivation. Comm. Fish. Rev. 10(9): 12-26

Nelson, R. C.
　1953.　Drill destruction by flame. Unpub. Rept. Elsworth Venus Res. Proj., J. & J. W. Elsworth Co., N. Y.

Nelson, T. C.
　1922.　The oyster drill: a brief account of their life history and possible means of combating them. Appendix Rept. Biol., N. J. Sta. Bd. Shellf., Nov., 1922.

　1923.　Report of the biologist. N. J. Agr. Exp. Sta. Ann. Rept. 1922: 335

　1939-
　1940.　Oysters, in Annual Report, N. J. Agr. Exp. Sta., 1939-40.

Newcombe, C. L. and staff.
　1941-
　1942.　Preliminary results of drill studies at Yorktown, Summary of 1941; and notes on drill trapping experiments conducted during the period June 3-Sept. 18, 1942. Unpub. Rept., Virginia Fish. Lab., Va.

Newcombe, C. L., and R. W. Menzel.
　1945.　Future of the Virginia oyster industry. Commonwealth 12(4): 3-11.

Orton, J. H.
    1909.   On the occurrence of protandric hermaphroditism in the mollusc Crepidula fornicata. Proc. Roy. Soc., Lond. 91B: 468-484.

    1927.   The habits and economic importance of the rough whelk-tingle (Murex erinaceus). Nature, Lond. 120: 653-655.

    1930.   On the oyster drill in the Essex estuaries. Essex Nat. 22: 298-306.

    1932.   The cold spring of 1929 in the British Isles. iii. Effect upon some marine animals in the oyster beds in the Thames Estuary. Acta Phaenol. s-Gravenhage 1: 129-132

Orton, J. H., and H. Mable Lewis.
    1931.   On the effect of the severe winter of 1928-29 on the oyster drills of the Blackwater Estuary. Jr. Mar. Biol. Assoc. U. K. 17: 3-1-313.

Orton, J. H., and R. Winckworth.
    1928.   The occurrence of the American oyster pest Urosalpinx cinerea (Say) on English oyster beds. Nature, Lond. 122: 241.

Owen, H. M.
    1947.   Observations on oyster drills: chromosomes of Urosalpinx cinereus Say. Conv. Addr., Nat. Shellf. Assoc., Asbury Park, N. J. (abstract).

Packard, E. L.
    1918.   Molluscan fauna from San Francisco Bay. Univ. Calif. Publ. Zool. 14: 200-457.

Pearse, A. S., and L. G. Williams.
    1951.   The biota of the reefs off the Carolinas. Jr. Elisha Mitchell Sci. Soc. 67: 133-161.

Pilsbry, H. A.
    1895.   Catalogue of the marine mollusks of Japan. Frederick Sterns, Detroit.

Pope, T. E. B.
    1910-   The oyster drill and other predatory Mollusca. Unpub.
    1911.   Rept. U. S. Bur. Fish., Washington, D. C.

Rathbun, R
      1888.   Report upon the inquiry respecting food fishes and the
              fishing grounds. U. S. Comm. Fish & Fish., Rept. Comm.
              1888(1892): XLI-CXXI.

Richards, H. G.
      1933.   Marine fossils from New Jersey indicating a mild inter-
              glacial stage. Proc. Amer. Philos. Soc. 72: 181-214

      1936.   Fauna of the Pleistocene Pamlico Formation of the southern
              Atlantic Coastal Plain. Bull. Geol. Soc. Amer. 47: 1611-1656.

      1938.   Marine Pleistocene of Florida. Bull. Geol. Soc. Amer. 49:
              1267-1296.

      1939.   Marine Pleistocene of the Gulf Coastal Plain: Alabama,
              Mississippi, and Louisiana. Bull. Geol. Soc. Amer. 50:
              297-316

      1939.   Marine Pleistocene of Texas. Bull. Geol. Soc. Amer. 50:
              1885-1898.

      1944.   Notes on the geology and paleontology of the Cape May Canal,
              New Jersey. Acad. Nat. Sci. Phila., Notulae Naturae, No. 134.

      1947.   Invertebrate fossils from deep wells along the Atlantic Coastal
              Plain. Jr. Paleontol. 21: 23-37.

      1950.   Geology of the Coastal Plain of North Carolina. Trans. Amer.
              Philos. Soc. 40: 1-83.

Rogers, Julia E.
      1951.   The Shell Book. Charles T. Branford Co., Boston, Mass.

Roughley, T. C.
      1925.   The story of the oyster. Its history, growth, cultivation, and
              pests in New South Wales. Australian Mus. Mag. 2: 1-32.

Rowe, H. C.
      1894.   Deep-water oyster culture. Bull. U.S. Fish Comm. 13: 273-276.

Ruge, J. G.
      1898.   The oysters and oyster beds of Florida. Bull. U.S. Fish Comm.
              for 1897: 289-296.

Ryder, J. A.
    1883.   Rearing oysters from artificially fertilized eggs, together
             with notes on pond culture. Bull. U.S. Fish Comm. 3: 281-294

Say, T.
    1822.   An account of some marine shells of the U. S. Jr. Acad.
             Nat. Sci. Phila. 2: 221-248.

Sherwood, H. P
    1931.   The oyster industry in North America: a record of a brief
             tour of some of the centers of the Atlantic and Pacific coasts
             and of a summer in Canada. Jr. du Conseil 6: 361-386.

Shimer, H. W., and R. R. Shrock.
    1944.   Index fossils of North America. Wiley, N. Y.

Sizer, I. W.
    1936.   Observations on the oyster drill with special reference to its
             movement and to the permeability of its egg case membrane.
             Unpub. Rept., U. S. Bur. Fish., Washington, D. C.

Smith, H. M.
    1907.   Our fish immigrants. Nat. Geogr. Mag. 18: 385-400.

Stauber, L. A.
    1938.   Oyster drill control in Delaware Bay. Conv. Addr., Nat.
             Shellf. Assoc., Providence, R. I.

    1941.   Survival of the oyster drill in waters of low salinity.
             Conv. Addr., Nat. Shellf. Assoc., Atlantic City, N.J.

    1941a.   The polyclad, Hoploplana inquilina thaisana Pearse, 1938,
             from the mantle cavity of oyster drills. Jr. Parasitol. 27:
             541-542.

    1943.   Ecological studies on the oyster drill, Urosalpinx cinerea,
             in Delaware Bay, with notes on the associated drill, Eupleura
             caudata, and with practical consideration of control methods.
             Unpub. Rept., Oyster Res. Lab., N.J.

    1943a.   Graphic representation of salinity in a tidal estuary.
             Jr. Mar. Res. 5(2): 165-167.

147

Stauber, L. A.
   1950.   The problem of physiological species with special reference
           to oysters and oyster drills. Ecology 31: 109-118.

Stauber, L. A., and G. M. Lehmuth.
   1937.   Costs and profits in oyster drill control.   Bull. N. J.
           Agr. Exp. Sta. 624: 1-10.

Stearns, R. E. C.
   1900.   Exotic mollusca in California.   Science, N. S. 11: 655-659.

Stimpson, W.
   1865.   On certain genera and families of zoophagous gastropods.
           Amer. Jr. Conchol. 1865: 55-64.

Storer, T.
   1931.   Known and potential results of bird and animal introduction with
           special reference to California. Mon. Bull. Dept. Agr. Sta.
           Calif. 20:267-273.

Stunkard, H. W , and C. R. Shaw.
   1931.   The effect of dilution on the activity and longevity of certain
           marine cercariae, with descriptions of two new species.
           Biol. Bull. 61: 242-271.

Suehiro, Y.
   1947.   A method of exterminating Rapana, a natural enemy of the
           oyster.   Bull. Jap. Soc. Sci. Fish. 13: 98-100.

Tarr, R. S.
   1885.   Natural enemies of oysters.   Science 6: 392.

Schenk, E. T., and J. H. McMasters
   1948.   Procedure in Taxonomy.   Stanford Univ. Press, Calif.

Townsend, C. H.
   1893. Report of observations respecting the oyster resources and
          oyster fishery of the Pacific coast of the United States. Rept.
          U. S. Comm. Fish & Fish. 1889-1891: 343-372.

Turner, H. J.
   1953.   The drilling mechanisms of the Naticidae.   Ecology 34(1):
           222-223.

Tryon, G. W.
    1873-
    1874.   American Marine Conchology. The Author, Phila., 208 pp.

    1882.   Structural and Systematic Conchology. The Author, Phila.

Uhler, P. R.
    1878.   List of animals observed at Fort Wool, Virginia. Johns
             Hopkins Univ., Sci. Results Chesapeake Zool. Lab. 1878: 17-34.

Verrill, A. E.
    1902.   The Bermuda Islands. Trans. Conn. Acad. Arts & Sci. 11:
             413-956

Verrill, A. E., and S. I. Smith.
    1874.   Report upon the Invertebrate Animals of Vineyard Sound and
             Adjacent Waters, with an Account of the Physical Features of the
             Region. Rept. Comm. Fish & Fish., Washington, D. C.

Walter, H. E.
    1910.   Variations in Urosalpinx. Amer. Nat. 44: 577-594.

Weeks, W. H.
    1907.   A collecting trip at Northport, N. Y. Nautilus 22: 98-99

Wentworth, E. P.
    1895.   Marine shells of the coast of Maine. Nautilus 9: 34-35.

Whiteaves, J. F
    1901.   Catalogue of the marine invertebrata of eastern Canada.
             Geol. Surv. Canada, Publ. 722.

Winkley, H. W.
    1912.   Notes on Maine Mollusca. Nautilus 26: 73-74.

Woelke, C. E
    1954.   Japanese oyster seed export program for 1954. Rept. Sta.
             Wash. Dept. Fish., Sta. Shellf. Lab., Quilcene, Wash.

Wood, A.E., and H. E. Wood.
    1927.   A quantitative study of the marine mollusks of Cape May
             County, New Jersey. Nautilus 41: 8-18.

Yonge, C. M.
1947.   The pallial organs in the aspidobranch gastropoda and
their evolution throughout the Mollusca.   Philos. Trans.
Roy. Soc. Lond., Ser. B., Biol. Sci., No. 591, Vol. 232:
443-518.